Trigonometry Geometry and The Conception of Space

Expanded First Edition

By Paul M Tokorcheck
University of Southern California

Bassim Hamadeh, CEO and Publisher
Michael Simpson, Vice President of Acquisitions
Jamie Giganti, Senior Managing Editor
Jess Busch, Senior Graphic Designer
Mark Combes, Senior Field Acquisitions Editor
Natalie Lakosil, Licensing Manager

Copyright © 2016 by Cognella, Inc. All rights reserved. No part of this publication may be reprinted, reproduced, transmitted, or utilized in any form or by any electronic, mechanical, or other means, now known or hereafter invented, including photocopying, microfilming, and recording, or in any information retrieval system without the written permission of Cognella, Inc.

First published in the United States of America in 2016 by Cognella, Inc.

Trademark Notice: Product or corporate names may be trademarks or registered trademarks, and are used only for identification and explanation without intent to infringe.

Printed in the United States of America

ISBN: 978-1-63487-187-7 (pbk) / 978-1-63487-188-4 (br)

For Tami.

We get to choose our friends, but not our family.

She's the best of both.

Preface

This text was first created to support a new course at Iowa State University, meeting for the first time in fall 2014. At the time this project was initiated, our standard trigonometry course was a prototypical pre-calculus offering, focusing on skills that were needed by STEM majors for their future calculus courses. However, the course rosters indicated that a large population of these students were actually from our College of Design. These students required a course that was more visual and more pragmatic, and that placed a greater emphasis on three-dimensional geometry.

At ISU this course asks for a standard course in college algebra as a prerequisite. General geometric knowledge (the Pythagorean Theorem), basic algebra skills (factoring), and a loose knowledge of functions and their graphs is all assumed. Students do not need to be experts in these topics; most are reviewed in this text to a certain extent. However, this book should not serve as an introduction to these topics in any capacity.

I would like to thank the faculty of Iowa State University for their support, not only for this project, but for all of the work I've done with them over the past few years. I'd also like to thank Mark Combes and Jamie Giganti of Cognella Academic Publishing for their help and guidance.

Last, but certainly not least, I am indebted to the staff of the Smokey Row Coffee Shop in Des Moines, Iowa. The bulk of this work was created at a corner table of that fine establishment, fueled by a steady supply of large lattes.

Paul Tokorcheck
Carlsbad, CA
June 2014

Contents

Sets and Numbers	1
Measurement and Angles	11
The Six Trigonometric Ratios	23
Standard Triangles	35
The Unit Circle	47
Functions and Inverse Functions	59
The Inverse Trigonometric Functions	71
Solving Equations for Angles	83
Using Trigonometric Identities	93
Applications of Trigonometry	105
The Law of Sines and the Law of Cosines	117
Coordinate Systems and Graphs	129
Polar Coordinates	143
Cylindrical Coordinates	153
Spherical Coordinates	165
Transformations of Graphs	175
Quadratic Equations and Conic Sections	187
Parabolas	197
Ellipses	207
Hyperbolas	221
Surfaces and Level Curves	233
Ellipsoids and Cones	243
Paraboloids	253
Hyperboloids	265

LESSON 1

Sets and Numbers

Our perception of the space around us is very much guided by our senses. A person who is blind in one eye cannot perceive depth and therefore must rely on other visual cues to determine the distance to an object. However, there is no denying that space has certain intrinsic properties that are not subject to our interpretation. For example, two distinct objects in a space have a positive, yet finite distance between them, and this distance is not dependent on our method of measurement, or even on our ability to perceive the distance in the first place.

We should note that the last statement described a distance as being both **positive** and **finite**.

If there is a chasm between the inherent properties of a space and our conceptualization of the space, then this statement already indicates the bridge. By associating a **number** to a given distance, we are able to compare one distance to another and create a framework in which we can discuss distances in general. This association is not trivial.

Nevertheless, we also make these associations with respect to more concrete objects. In modern language it's common to say that we "multiply two sides of a rectangle to find its area," though the side of a rectangle is geometric object, not a number. We cannot "multiply two sides" any more than we can multiply two refrigerators.

Lesson 1

What we mean by this statement is that we can assign a numerical value to indicate the **length** of each side, which is the distance from one end of the side to the other. We then multiply the two numbers, and interpret the product as another, entirely different, geometric quantity.

These may seem to be insignificant semantic distinctions. However, there are a variety of ways to measure quantities like length or area, which are sometimes in direct competition with each other. We'll not only need to distinguish between different types of quantities, but also the **units** we use when assigning a number to a quantity.

How to artfully **use** a space is the realm of architecture and interior design. But how to accurately **measure** the dimensions of that space, well, that is very much at the heart of mathematics and its history. This is a history that dates back thousands of years, from Bronze Age Sumerian farmers measuring their fields, and the astronomers of China and India, to the geometers of ancient Greece.

In later lessons we'll focus more on the idea of **position**, which will require some more modern tools. Position is not indicated by a single number, but by a larger set of data. Again, there are various ways to do this. We will learn how to identify points using several methods, and how to transfer between them. Before we do, we need to discuss the basic concepts of **numbers** and **sets**.

Definition: A **set** is a collection of objects, each of which is called an **element**. If x is an element of A, we write

$$x \in A.$$

To say that x is **not** an element of A, we write

$$x \notin A.$$

There are several ways to describe a set. One common method is simply to list the elements of the set within a pair of braces, as shown here:

$$A = \{\text{chair, lamp, sofa, table}\}$$

$$B = \{\text{red, blue, black, green, yellow}\}$$

Note that both of these sets contain a finite number of elements; A contains four elements, and B contains five.

Definition: The number of elements in a set is called its **cardinality**. The set with no elements is called **empty set**, or **null set**. It is denoted by $\emptyset$.

Of course, when a set contains many elements, listing them individually becomes impractical.

LESSON 1

One way to describe sets with many elements is to include an ellipsis, and leave it to the reader to discern the pattern that we are trying to convey. For example:

$$A = \{\text{bird, cat, fish, dog,} \ldots\}$$

$$B = \{\ldots, -5, -3, -1, 1, 3, 5, 7, \ldots\}$$

While commonly used, this method can unfortunately lead to ambiguity. Is set A above meant to include **all** animals? Just those kept as pets? The description of set B is perhaps a bit more more clear, but we are still relying on the reader to recognize that all of the listed numbers are odd, and that this is the defining feature of this set.

An even better method uses **set builder** notation. In this notation, an element is listed along with a short description of the properties that elements of the set should have. The name of the element and its decription are separated either by a vertical line or by a colon. Here are a few examples that may or may not be sets of numbers:

$$A = \{a \mid a \text{ is an even number}\}$$

$$B = \{b \in A : b > 5\}$$

$$C = \{\text{cars} \mid \text{made before 1995}\}$$

Definition: The **union** of two sets A and B is defined by

$$A \cup B = \{x \mid x \in A \text{ or } x \in B\}.$$

The **intersection** of two sets A and B is defined by

$$A \cap B = \{x \mid x \in A \text{ and } x \in B\}.$$

The **difference** of two sets A and B is defined by

$$A \setminus B = \{x \mid x \in A \text{ and } x \notin B\}.$$

In these definitions, note the use of the words "and" and "or". In mathematics, "or" is always intended to be **inclusive**: $x \in A \cup B$ if x is an element of A, if x is an element of B, or if x is an element of both. However, for x to be an element of the intersection, it must necessarily be an element of both A and B.

Example: Let A and B be the following two sets:

$$A = \{1, 24, -4, 7, 13, -17\},$$

$$B = \{0, 3, 6, 9, 12, \ldots\}.$$

Then $A \cap B = \{24\}$. It would be awkward to try to write $A \cup B$ within a single pair of set brackets, which is why we have the notation to help.

LESSON 1

In this course, we will be concerned primarily with sets of numbers, so we would be well served to define several common collections of numbers that we can refer to later.

The most basic set of numbers that we will consider is the set of **natural numbers**, which consists of the positive whole numbers. You might think of these as the numbers that describe the cardinality of a finite, non-empty set. In the notation of sets, we write the natural numbers as

$$\mathbb{N} = \{1, 2, 3, 4, 5, \dots\}.$$

Note that $12 \in \mathbb{N}$, but $-2 \notin \mathbb{N}$, because -2 is not positive, and there is no set with -2 elements. Likewise, $7.3 \notin \mathbb{N}$, because 7.3 is not a whole number.

We can extend $\mathbb{N}$ to a larger set by including the **negatives** of all of these natural numbers. We will also include the number **zero**, which has the distinguishing feature that it does not add anything via addition: for any number a,

$$a + 0 = a.$$

For this reason, we call **zero** the **additive identity**. The new set created by these additions is called the **integers**, and is denoted by the symbol $\mathbb{Z}$:

$$\mathbb{Z} = \{\dots, -3, -2, -1, 0, 1, 2, 3, \dots\}.$$

While the number -2 is an integer, the number 7.3 is not. If we imagine the natural numbers as those used to count objects, then the integers imply an understanding of deficit, in which some number of objects is missing or owed.

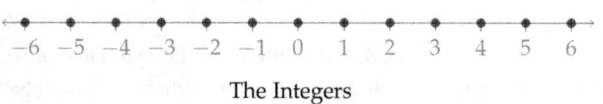

The Integers

In a more mathematical sense, we consider the negatives to be the **inverses** of the natural numbers. That is, if $a \in \mathbb{N}$, then $-a$ is the unique integer for which

$$a + (-a) = 0.$$

As we talk about identities and inverses, we note that there is another integer with a property similar to zero. Namely, the number 1 has the property that for any number a,

$$a \cdot 1 = a.$$

For this reason, we call **one** the **multiplicative identity**. However, here there aren't any inverses as there were in the additive sense, at least not within the integers.

Even for the next smallest integer, 2, there isn't any other integer $b \in \mathbb{Z}$ such that $2 \cdot b = 1$. In fact, we could show that if a is **any** integer that is not equal to 1, there cannot be another integer b for which $a \cdot b = 1$.

If we want a set that contains these multiplicative inverses, sometimes called **reciprocals**, then we'll need to construct it. The **rational numbers** are denoted by the symbol $\mathbb{Q}$, and defined to be

$$\mathbb{Q} = \left\{ \frac{a}{b} \;\middle|\; a, b \in \mathbb{Z},\ b \neq 0 \right\}.$$

Rational numbers are **ratios** of one integer to another, and the set includes all of the integers themselves, if we take the denominator b to be 1. They follow all of the standard rules of arithmetic for fractions, including multiplication,

$$\frac{a}{b} \cdot \frac{c}{d} = \frac{ac}{bd},$$

and addition using a common denominator,

$$\frac{a}{b} + \frac{c}{d} = \frac{ad + bc}{bd}.$$

The rational numbers include all terminating decimals, which can be written as ratios as in the following example:

$$5.193708 = \frac{5193708}{1000000}.$$

Similarly, many rational numbers have infinitely repeating decimals when divided, such as

$$\frac{2}{3} = 0.\overline{6} = 0.66666...$$

Some Rational Numbers

One property of rational numbers that distinguishes them from integers is easily demonstrated if we place each number on a number line. Namely, the rational numbers are infinitely **dense**, in the sense that in between every two rational numbers, there is another rational number.

Some (more) Rational Numbers

It may seem that the rational numbers will fill the line completely, so that every point on the line will have a rational number associated with it. However, it has been long known that this is not the case.

LESSON 1

As an ancient example, consider a square, whose sides are all of length one. Then, draw the **diagonal** of the square from one corner to another.

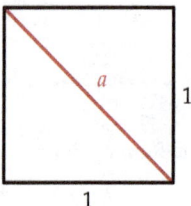

It is clear that the diagonal has some positive length a, and it is also clear that a is larger than 1. But what is a exactly?

This was a confusing problem from the perspective of our ancient cultures. For example, the Greeks were very familiar with rational numbers and their arithmetic. And though they would not have written it in our modern form, they were also familiar with **the Pythagorean Theorem**, which in this context tells us that a should satisfy the formula

$$a^2 = 1^2 + 1^2.$$

That is, whatever the length of a is, it should satisfy $a^2 = 2$. But attempting to find a rational number a that satisfies this equation quickly proved impossible.

We know that $1 < a$, and since $2^2 = 4$, we find that $a < 2$ as well. We might attempt something in between, such as

$$(1.5)^2 = \left(\frac{3}{2}\right)^2 = \frac{9}{4} = 2.25,$$

but this only shows us that $a < 1.5$. We might next try

$$(1.4)^2 = \left(\frac{14}{10}\right)^2 = \left(\frac{7}{5}\right)^2 = \frac{49}{25} = 1.96,$$

from which we conclude that $1.4 < a < 1.5$. Between these two numbers lies 1.45, which we can also test:

$$(1.45)^2 = \left(\frac{145}{100}\right)^2 = \frac{21025}{10000} = \frac{841}{400} = 2.1025.$$

This number is still too large, showing that $1.40 < a < 1.45$. Continuing this process would get us very **close** to a value of a, which can be estimated to 40 decimal places by

$$a \approx 1.4142135623730950488016887242096980785697.$$

However, there is no finite decimal or fraction that will ever square to be **equal** to 2.

And yet, there is very definitely a distance a, which is easily constructed geometrically. This leads to the conclusion that there are distances, or positions on the number line, that do not correspond to any rational number. If our method of understanding spatial relationships is to associate a number to a distance, this new fact may be somewhat disheartening.

What is worse, this same argument can be repeated for **any** square of any rational side length. Therefore, not only are there infinitely many of these **irrational numbers**, but there are even more of them than there are rational numbers. Not only are there distances on the number line that have no rational number associated to them, but in fact **most** distances are described by irrational numbers.

It's clear that the rational numbers $\mathbb{Q}$ are not sufficient for our purposes here, so we'll extend this set of numbers yet again. The **real numbers** are the set of distances on a line, and are denoted by $\mathbb{R}$.

It would be quite difficult to write the set $\mathbb{R}$ in any sort of set notation as we did for our previous number systems. However, we can say that $\mathbb{R}$ contains $\mathbb{Q}$, meaning that every rational number can be associated to a distance on a line. The sets $\mathbb{N}$ and $\mathbb{Z}$ are likewise contained in $\mathbb{R}$ as well.

But $\mathbb{R}$ also includes irrational numbers like our a, which had the property that $a^2 = 2$. As an element of $\mathbb{R}$ we give this number a name, the **square root** of 2, and denote it by $a = \sqrt{2}$. So long as k is a positive real number, we can make a similar definition of $\sqrt{k}$, and almost all of these square roots will be irrational.

In addition, there are many more irrational numbers that do not arise as square roots or any other type of root, including one irrational number that plays a central role in this course.

Definition: The ratio of the circumference of a circle to its diameter is a fixed constant, denoted by the Greek letter π:

$$\pi = \frac{c}{d}.$$

This ratio is an irrational number, which shows π as being approximately equal to 3.1415926536.

The line of real numbers is dense with irrational numbers in the same way that it was dense with rationals; in between any two irrational numbers is another one. These two types of numbers share the same space on the real line, infinitely densely mixed together, and together they **complete** the line into a single continuous geometric object with no holes.

The subsets of $\mathbb{R}$ that we have seen so far do not share this property of continuity, but there are other subsets of the real line that do, which we describe next.

LESSON 1

Definition: An **interval** is a subset of the real line bounded by two endpoints. We denote intervals as:

$$[a,b] = \{x \in \mathbb{R} \mid a \leq x \leq b\}$$
$$(a,b) = \{x \in \mathbb{R} \mid a < x < b\}$$
$$(a,b] = \{x \in \mathbb{R} \mid a < x \leq b\}$$
$$[a,b) = \{x \in \mathbb{R} \mid a \leq x < b\}$$

Intervals $[a,b]$ that contain their endpoints are called **closed**. Intervals (a,b) that do not contain their endpoints are called **open**.

Definition: For intervals with only one endpoint, we use the infinity symbol ∞ for the open end of the interval:

$$[a,\infty) = \{x \in \mathbb{R} \mid a \leq x\}$$
$$(a,\infty) = \{x \in \mathbb{R} \mid a < x\}$$
$$(-\infty,b] = \{x \in \mathbb{R} \mid x \leq b\}$$
$$(-\infty,b) = \{x \in \mathbb{R} \mid x < b\}$$

Often two or more of these intervals are combined to make a larger one. Or, sometimes two intervals will overlap, and we will want to identify where that happens.

Example: Let S denote the interval $(-\infty, -3)$, and let T denote the interval $[1, \infty)$:

$$S = \{x \in \mathbb{R} \mid -3 > x\},$$
$$T = \{x \in \mathbb{R} \mid x \geq 1\}.$$

We can visualize these subsets of the number line by shading in the intervals represented by the two sets. We shade the interval S in below in blue, and T in red. Note the difference between the two endpoints. For T, we color the point $x = 1$ with a solid dot, which indicates that it is part of the interval. For S, we draw the point $x = -3$ with an open circle, indicating that it is not **part** of the interval.

From the image it is clear that S and T have no points in common, so $S \cap T = \emptyset$. In set-builder notation, we might write the union of these sets as

$$S \cup T = \{x \in \mathbb{R} \mid -3 > x \text{ and } x \geq 1\}.$$

But as an interval we can only write the union literally:

$$S \cup T = (-\infty, -3) \cup [1, \infty).$$

Since the two intervals have no intersection, their set differences are somewhat trivial:

$$S \setminus T = S, \quad T \setminus S = T.$$

To see some set differences that are a bit more interesting, we'll look at another pair of intervals.

Example: Let S denote the interval $(-2, \infty)$, and let T denote the interval $(4, \infty]$:

$$S = \{x \in \mathbb{R} \mid x > -2\},$$
$$T = \{x \in \mathbb{R} \mid x \geq 4\}.$$

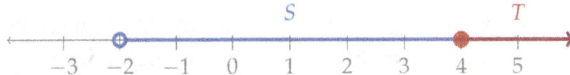

Here every one of the points of T are also in S, so removing all points of S from T will leaves nothing: $T \setminus S = \emptyset$.

Conversely, removing all elements of T from S leaves us with the set of (strictly) blue points on the number line. We must pay careful attention to the endpoints. The point $x = 4$ belongs to T, and therefore does **not** belong to $S \setminus T$:

$$S \setminus T = (-2, 4) = \{x \in \mathbb{R} \mid -2 < x < 4\}.$$

Example: Find the solutions to the following inequality, and write the set of solutions in interval notation.

$$2(3n+4) - 2 \leq 3(1+n).$$

This type of problem serves as a review of general algebra skills, along with some practice writing sets and intervals. Many operations could be carried out first, but we will choose to distribute the constants into both sets of brackets:

$$2(3n+4) - 2 \leq 3(1+n)$$
$$(6n+8) - 2 \leq (3+3n).$$

Now we'll combine the constant terms on the left side:

$$6n + 6 \leq 3 + 3n.$$

Subtracting 3 from both sides gives:

$$6n + 3 \leq -3n,$$

and subtracting $6n$ from both sides gives:

$$3 \leq -3n.$$

We'll now divide both sides of this equation by -3. Recall that since we are dividing by a **negative** number, we must switch the direction of the inequality:

$$-1 \geq n.$$

In interval notation, this is written $(-\infty, -1]$.

LESSON 1

EXERCISES

For Exercises 1–8, let O denote the set of odd numbers, let E denote the set of even numbers, and let $A = \{1,4,5,7,8\}$. Describe the following sets:

1. $A \cap E$
2. $A \cap O$
3. $A \setminus E$
4. $E \cap O$
5. $O \setminus A$
6. $E \cup O$
7. $E \setminus A$
8. $(A \cap O) \cup E$

For Exercises 9–12, determine the elements of each set that are (a) natural numbers, (b) integers, (c) rational numbers, and (d) irrational numbers.

9. $R = \left\{ 8,\ 2.05,\ \dfrac{9}{4},\ -5,\ \sqrt{7} \right\}$
10. $S = \left\{ 0.\overline{6},\ -\dfrac{2}{3},\ 0,\ 2\sqrt{3} \right\}$
11. $T = \{6.5,\ 1.010110111\ldots\}$
12. $U = \left\{ \dfrac{\pi}{2},\ \sqrt{16},\ 5^{3/2},\ 1 \right\}$

For Exercises 13–16, find the distance between a and b on the number line.

13. $a = \dfrac{1}{2},\ b = -5$
14. $a = 4,\ b = 13$
15. $a = -\dfrac{3}{5},\ b = \dfrac{2}{5}$
16. $a = \pi,\ b = 5\pi$

17. Let A be the set of real numbers that are greater than 4, but less than or equal to 15. Write A in interval notation.

18. Let B be the set of real numbers that are greater than or equal to -5. Write B in interval notation.

For Exercises 19–22, use a calculator to estimate each irrational number. Round to five decimal places.

19. $\dfrac{\sqrt{3}}{2}$
20. 2π
21. $\dfrac{\sqrt{2}}{2}$
22. π^2

For Exercises 23–26, solve each inequality. Draw the solution set on a number line, and write it in interval notation.

23. $5x + 4 \geq 34$
24. $9(k+2) < 72$
25. $4(4z+5) - 5 > 3(4z-1)$
26. $\dfrac{2x+3}{5} \leq 0.03$

27. List the smallest four elements of the set
$$S = \{y \in \mathbb{Z} \mid y = x^2 - 1\}.$$

28. List the smallest four elements of the set
$$S = \{z \mid z = |x| - x,\ x \text{ is a negative integer}\}.$$

LESSON 2

Measurement and Angles

In Lesson 1 we explored some different sets of numbers and their properties, along with the idea of associating real numbers to geometric objects. Specifically, we focused on real numbers as distances or lengths. However, lengths are not the only physical attribute that can be measured. In all cases the act of measurement consists of making such an association, and here we'll look at this process more carefully.

The first notion we'll discuss is that of **dimension**. In our current usage the word "dimension" will not refer to the three spatial dimensions, though we will talk about two- and three-dimensional spaces later on. Here, you might imagine talking about "the dimensions of a basketball court," referring not to the fact that the court is two-dimensional, but rather to the measurable **shape** and **size** of the court.

The current SI system of measurement recognizes seven basic physical dimensions: **mass** (M), **length** (L), **time** (T), **electrical current** (I), **luminous intensity** (J), **amount of substance** (N), and **temperature** (Θ).

So, here the word "dimension" is used to create a distinction between different measurable attributes. Only quantities of the same dimension may be added together: we cannot add a length to a temperature, for example, as they have different dimensions.

Lesson 2

However, we **can** multiply or divide any of these basic dimensions, and in this way we obtain new dimensions for other measurable quantities that are not listed. A **volume** can be acquired by multiplying three lengths, and a **speed** can be acquired by dividing a length by a time. Speed is now a new dimension, and we cannot add together a speed and a length to obtain anything meaningful.

In order to assign a real number to an attribute of any dimension, we introduce **units of measure**. Dimensions and units may seem to be logically the same, but they are actually fundamentally different ideas. For example, mass is a dimension, but the **kilogram** is a unit that we use to measure mass. These units are human-made standards that are agreed upon by convention, though there are often several such standards in place for any given dimension. Changing between units is often a valid problem in its own right.

Using length as an example, let's take two quantities of the same dimension and also the same unit, like $a = 4$ inches and $b = 28$ inches. Multiplying these two quantities gives

$$ab = (4 \text{ in})(b = 28 \text{ in}) = (4)(28) \text{ in}^2 = 108 \text{ in}^2.$$

We see that this new quantity is of a different dimension than either of the originals; it represents an **area**.

But consider what happens if we divide instead:

$$\frac{a}{b} = \frac{4 \text{ in}}{28 \text{ in}} = \frac{1 \text{ in}}{7 \text{ in}} = \frac{1}{7}.$$

Instead of creating a new dimension, this creates a dimensionless, unitless **ratio** of the two values. The ratio loses all reference to the original units. Indeed, the ratio of 4 km to 28 km will be exactly the same, as will the ratio of 5 grams to 35 grams.

These ratios give us a way to convert from one unit to another of the same dimension. Since 1 foot is equal to 12 inches, the ratio of one to the other is exactly one:

$$\frac{1 \text{ foot}}{12 \text{ inches}} = \frac{12 \text{ inches}}{12 \text{ inches}} = \frac{12}{12} = 1.$$

Again, this is a ratio of two units of the same dimension, and is itself unitless and dimensionless. Therefore, multiplying it by any other quantity leaves that quantity unchanged, both numerically and in terms of its dimension. So if we wished to convert $b = 28$ inches into feet, we multiply by this ratio is a specific way:

$$b = 28 \text{ inches} = 28 \text{ in} \cdot \frac{1 \text{ ft}}{12 \text{ in}} = \frac{28 \text{ ft} \cdot \text{in}}{12 \text{ in}} = \frac{7}{3} \text{ ft} = 2.\overline{3} \text{ ft}.$$

Notice the way that the conversion ratio was written. The unit we wish to convert from is in the denominator, ensuring that it will cancel with the unit we already have. The unit we are converting to is in the numerator, and since it does not appear anywhere else in the expression, it will remain there throughout the calculation.

Example: One mile is equal to 1.60934 kilometers. The driving distance from Des Moines to Ames, Iowa is 37.5 miles. How many kilometers is this?

Our conversion ratio will be a fraction involving 1 mile and 1.60934 kilometers. Since the two are equal, this could take either of the forms

$$\frac{1.60934 \text{ km}}{1 \text{ mile}} \quad \text{or} \quad \frac{1 \text{ mile}}{1.60934 \text{ km}}.$$

We would like to end with units of kilometers, this will be in the numerator of the fraction:

$$37.5 \text{ mi} = 37.5 \, \cancel{\text{mi}} \cdot \frac{1.60934 \text{ km}}{1 \, \cancel{\text{mi}}} = 60.35025 \text{ km}.$$

This example illustrates some of the algebraic rules that come into play when working with units, which we can summarize in a few bullet points.

The manipulation of these units is sometimes referred to **dimensional analysis**.

- Only units of the same dimension may be added or subtracted. Therefore any differing units should be converted to a single, common unit before adding the numerical quantities.

- Units of any dimension may be multiplied or divided. When doing so, the units should be treated as algebraic variables, using the laws of exponents for multiplication and division. Different units should be treated as unique variables, even if they have the same dimension.

- To convert a measured quantity from one unit to another unit of the same dimension, find two quantities of each unit that are equal to each other, and write them together as a fraction or ratio. Then multiply or divide your measured quantity by this ratio so that the original units cancel.

In the next example we explore a question that is not a simple conversion from one unit to another. Instead, we are given multiple pieces of information in a variety of units, and asked to identify a measure of time.

LESSON 2

Example: A molecule of hydrogen moves in a straight line at a speed of 115 cm/s. How long will it take to travel the length of a football field, 100 yards? (One yard is equal to 0.9144 meters.)

We first begin with the two given quantities: the speed of the molecule and the length of the field. These two attributes have different dimensions, so we cannot add them. We can multiply them, however, and we would like to do so in such a way that the result will be a quantity of time.

Multiplying these directly will give

$$115 \, \frac{\text{cm}}{\text{s}} \cdot 100 \text{ yd} = 11{,}500 \, \frac{\text{cm} \cdot \text{yd}}{\text{s}}.$$

We see that the units of length don't cancel as we would like. Also, the unit of time (seconds) is in the denominator, though we would like it to end in the numerator. We can fix both of these problems by inverting the first term:

$$\frac{1 \text{ s}}{115 \text{ cm}} \cdot 100 \text{ yd} = \frac{100}{115} \, \frac{\text{s} \cdot \text{yd}}{\text{cm}}.$$

This works much better, at least in terms of the dimensions. But we find that the units of length still do not cancel; they measure the same dimension, but in different units.

We will need to convert them to a common unit in order to cancel them. The hint in the question helps us a bit, giving us one conversion ratio:

$$\frac{1 \text{ s}}{115 \text{ cm}} \cdot 100 \text{ yd} \cdot \frac{0.9144 \text{ m}}{1 \text{ yd}}.$$

The units have almost worked themselves out, but we still need a relationship between meters and centimeters. In the SI standard of prefixes, 1 meter is divided into 100 centimeters, and this provides us with our next conversion ratio. Canceling out the units and grouping together the numerical terms, we find that

$$\frac{1 \text{ s}}{115 \text{ cm}} \cdot 100 \text{ yd} \cdot \frac{0.9144 \text{ m}}{1 \text{ yd}} \cdot \frac{100 \text{ cm}}{1 \text{ m}} = \frac{9144}{115} \text{ s} \approx 79.513 \text{ s}.$$

The methods used in this last example are very common. We first identify the dimension and units that we would like our final answer to be in. The next step generally involves some amount of trial and error, trying to write each piece of information so that the unwanted units cancel.

We will practice these methods of dimensional analysis for the remainder of this lesson as we measure **angles**. You will note that angles do not fall within the seven basic dimensions recognized by the SI system of measurement.

Instead, we will have to define angles ourselves and construct a unit for them in terms of the standard units we already have.

Definition: An **angle** is a geometric object, formed by two lines or line segments that intersect at a common point. The two lines or line segments are called the **sides** of the angle. The single point at which they intersect is called the **vertex** of the angle.

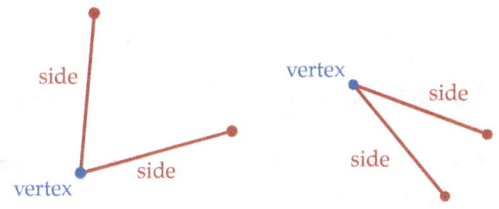

It seems clear from the above image that one of these angles is "larger" than the other, but we need a way to make this statement precise. One way of doing this uses the length of a **circular arc** that passes from one line segment to the other. We can draw such an arc using either a piece of string of a fixed length r_1, or a compass set so that the tips of its arms are distance r_1 apart. This length will be called the **radius** of resulting arc.

If we place one end of the compass at the vertex, and rotate the other end from one side of the angle to the other, we form a circular arc spanning the interior of the angle. This arc will have a length l_1 of its own.

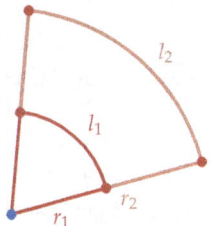

What is more, if we choose another radius r_2 and construct the arc of length l_2 in exactly the same way, then the **ratio** of l_1 to r_1 will be identical to the ratio of l_2 to r_2:

$$\frac{l_1}{r_1} = \frac{l_2}{r_2}.$$

Since this quantity is unique to the angle, and does not depend on the radius chosen to create it, we can use this real number as the **measure** of our angle. It is worth noting, however, that as a ratio of two lengths, this is a **dimensionless quantity**, like our earlier conversion ratios. If you like, you can think of an angle as a conversion ratio between linear length and arc length.

Lesson 2

Definition: Suppose that a fixed angle is given. For any chosen radius, a circular arc may be drawn from one side of the angle to the other. The ratio of the length of this arc to the radius will be a fixed constant. This ratio defines the measure of the angle.

$$\theta = \frac{\text{length of the arc}}{\text{length of the radius used to create the arc}}.$$

We commonly refer to unknown angles with lowercase Greek letters, such as α, θ, or ϕ. An exception is the letter π which is reserved as a fixed irrational constant.

Example: Suppose that the endpoints of two line segments meet at a vertex and that both segments have length $r = 1$ cm. The segments together create a single straight line with the two line segments parallel to each other, as shown in the image below.

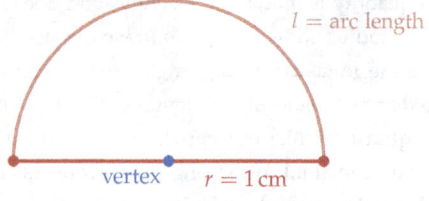

This angle and radius r create an arc length l. To find the value of l, and therefore the measure of our angle, we recall our definition of the irrational number π. If C is the circumference of a circle, and d is the diameter, then

$$\pi = \frac{C}{d}.$$

But in our case, the arc length l is exactly half of the circumference of a full circle, or $C = 2l$. Also, the diameter d is twice our radius, $d = 2r$, so substituting we have

$$\pi = \frac{2l}{2r} = \frac{l}{r}.$$

What we find is that the measure of our angle is exactly π.

Once again, as the ratio of two lengths, this is a dimensionless quantity. Nevertheless, we may still define a standard **unit** of measurement for angles measured in this way, called a **radian**.

Example: An angle α is drawn on a piece of paper. A compass is set so that its radius is 4 cm. When the compass is used to draw a circular arc from one side of α to the other, we use a string to find that the length of the arc is 5 cm. We conclude that

$$\alpha = \frac{5 \text{ cm}}{4 \text{ cm}} = 1.2 \text{ rad}.$$

Example: Conversely, suppose that an angle $\beta = 3.5$ radians is drawn, and we use a compass set to a radius of 2 cm to draw a circular arc from one side of β to the other. We can find the length of the circular arc using the definition of radian measure:

$$3.5 \text{ rad} = \frac{\text{arc length}}{2 \text{ cm}}$$

$$(3.5 \text{ rad})(2 \text{ cm}) = \text{arc length}$$

$$\text{arc length} = 7 \text{ cm}.$$

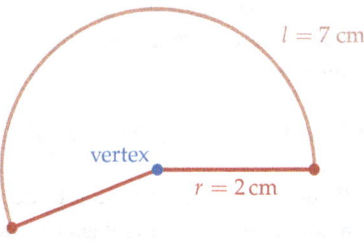

Note how the units interact with each other in this example. Because radians are defined as a dimensionless ratio, the multiplication of a length by an angle will still be a measurement of length. Again, we can consider an angle to be a conversion ratio between linear length and arc length.

We've now seen two different angles, measuring π radians and 3.5 radians, respectively. Drawing these angles can help us get an intuition for how "big" a radian is, in terms of angle. More precisely, each segment of angle in the image below measures exactly one radian.

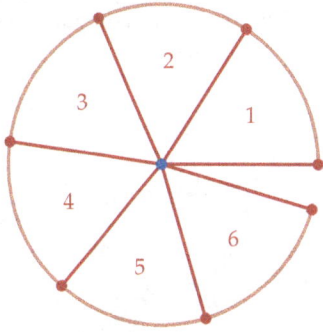

You'll notice that a full circle consists of slightly more than six full radians. We should expect this, since we already know that **half** of a circle consists of $\pi \approx 3.14$ radians. More exactly, a circle of radius 1 cm has a circumference of 2π cm, and therefore the angle measure is of a full circle is

$$\theta = \frac{2\pi \text{ cm}}{1 \text{ cm}} = 2\pi \text{ rad}.$$

LESSON 2

We will return to radian measure very shortly, but first we'll introduce an alternative measurement of angle. While radians have some very nice mathematical properties, the fact that most common angles have irrational measures might persuade us to try something else, where the arithmetic is not quite so cumbersome.

Definition: A full circle can be divided into 360 equal parts, which we call **degrees** (°).

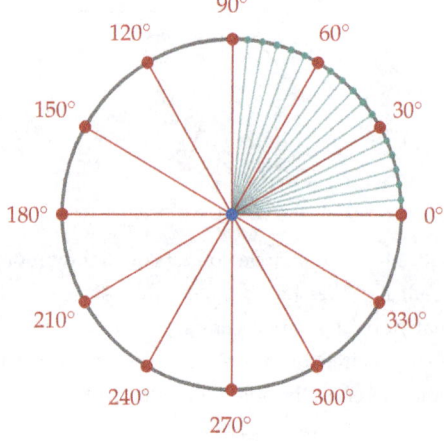

For finer measurements of degrees, we could use a decimal, as we often do with radians. Alternatively, we can further divide each degree into 60 **minutes** ($'$), and each minute into 60 **seconds** ($''$). This gives us a collection of different units of varying size to use for angles.

However, regardless of the units that we choose to use for measuring angles, the angle measure is always a dimensionless quantity, representing a ratio of two lengths.

Example: Convert $47°36'52''$ to a pure degree measurement, written as a decimal.

We first write the given angle as a sum of its individual parts, each with its own unit.

$$47°36'52'' = 47° + 36' + 52''.$$

Then, we will convert each of the parts to degrees using dimensional analysis, before adding them back together:

$$47°36'52'' = 47° + 36' \cdot \frac{1°}{60'} + 52'' \cdot \frac{1°}{60'} \cdot \frac{1'}{60''}$$

$$= 47° + 0.6° + 0.01\overline{4}°$$

$$= 47.61\overline{4}°.$$

Example: Convert the angle $\theta = 132.8525°$ into degrees, minutes, and seconds.

We try reversing our steps from the above example. Separating off the $132°$ that is the whole part of the degree measurement, we consider the decimal remainder $0.8525°$. We convert it from degrees to minutes, writing out the full units for clarity:

$$0.8525 \text{ degrees} \cdot \frac{60 \text{ minutes}}{1 \text{ degree}} = 51.15 \text{ minutes.}$$

Now we have that $\theta = 132° + 51.15'$. Next, separate off the decimal part of this, $0.15'$, and convert it to seconds:

$$0.15 \text{ minutes} \cdot \frac{60 \text{ seconds}}{1 \text{ minute}} = 9 \text{ seconds.}$$

Now we have a final conversion of

$$132.8525° = 132 + 0.85 + 0.0025 \text{ degrees} = 132°51'9''.$$

We now have two sets of units with which to measure angles. Both of them have their advantages and disadvantages, and we will be using both radians and degrees throughout this book, and therefore we will be converting units of angle very often.

To find a connection between radians and degrees, we look to our most natural, built-in unit of angle, the full circle:

$$1 \text{ full circle} = 360° = 2\pi \text{ rad.}$$

Having this equality between our two standard units of angle, we can now use it to perform conversions between them. As a first example, how many degrees would make an angle of 1 radian? We use dimensional analysis to check:

$$1 \text{ rad} \cdot \frac{360 \text{ degrees}}{2\pi \text{ rad}} = \frac{180}{\pi} \text{ degrees} \cong 57.29578°.$$

Example: Convert the angle $90°$ into radian measure.

Again, we use dimensional analysis. Note the way that the converting fraction is written, so that the degree units will cancel with each other.

$$90 \text{ degrees} \cdot \frac{2\pi \text{ rad}}{360 \text{ degrees}} = \frac{\pi}{2} \text{ rad} \cong 1.5708 \text{ rad.}$$

This should be consistent with our previous experience. Earlier we found that **half** of a full circle measures π radians (or $180°$), so we should expect the measure of a **quarter** circle to be exactly half of that, or a quarter of the full 2π radian circle.

LESSON 2

We can use these conversions to draw a few standard angles, and write them using both units for comparison.

When drawing angles in a two-dimensional plane, such as a piece of paper or a chalkboard, it is customary to form them in the **counterclockwise** direction. We have done this below, measuring from the side labeled $\theta = 0°$.

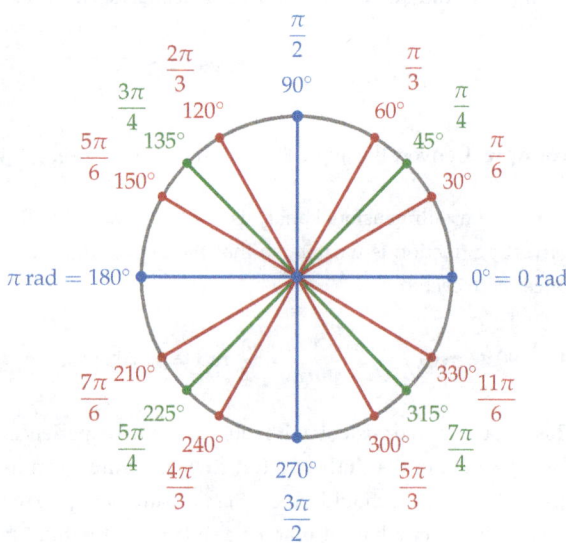

Using this conventional orientation makes it possible to distinguish between the two sides.

Definition: The side of an angle in which the angle originates is called the **initial** side, and the side in which it terminates is called the **terminal** side.

Furthermore, once the initial side of an angle is fixed, we can make logical sense of angles spanning more than 360° or 2π radians. Below we draw the angle $\theta = 510°$, which completes more than one full circle:

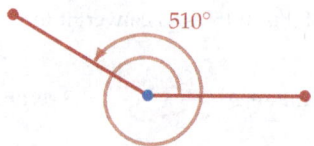

If we subtract a full circle's measure of 360°, we find that the remaining, smallest angle between these two line segments is $510° - 360° = 150°$.

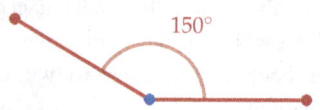

20

LESSON 2

Angles were defined, as a geometric objects, to be a pair of line segments that intersect at a common point. So in this sense, the two angles 510° and 150° might be considered to be "equal." But we sometimes want to emphasize that two angles are **not** equal, but simply have terminal sides that point in the same direction. In this case we use the following adjective.

Definition: Two angles α and β that have the same measure are called **equal** or **congruent**. Two angles are called **coterminal** if their difference is a multiple of 2π (or 360°).

We'll try to motivate the above definition with an example involving arc length.

Example: Suppose that a circular arc is drawn with radius $r = 6$ in, through an angle of $\theta = 715°$. Find the length of the arc, estimated to five decimal places.

In this example, θ is given with its units in degrees. However, circular arc lengths are much more natural to calculate with radians, as we've seen in the previous examples. Therefore we'll first convert this angle to radians:

$$\theta = 715° \cdot \frac{2\pi \text{ rad}}{360°} = \frac{143\pi}{36} \text{ rad}.$$

Now we can use this in the definition of radian measure:

$$\frac{\text{arc length}}{\text{radius}} = \theta$$

$$\frac{\text{arc length}}{6 \text{ in}} = \frac{143\pi}{36} \text{ rad}$$

$$\text{arc length} = \left(\frac{143\pi}{36} \text{ rad}\right)(6 \text{ in})$$

$$\text{arc length} = \frac{143\pi}{6} \text{ in}$$

$$\text{arc length} \approx 74.87462 \text{ in.}$$

However, the angle $\theta = 715°$ is larger than one complete revolution. It is coterminal to

$$\phi = 715° - 360° = 355°,$$

but using $\phi = 355°$ in the above calculation would give us a very different arc length:

$$\text{arc length} = 355° \cdot \frac{2\pi \text{ rad}}{360°} \cdot 6 \text{ in} = \frac{71\pi}{6} \text{ in} \approx 37.175513 \text{ in.}$$

Because of this, we'll make every attempt to distinguish between angles that are not equal, but are simply coterminal.

LESSON 2

EXERCISES

1. How many minutes will it take a car traveling 75 miles per hour to travel a distance of 123 miles?

2. How many feet are in $\frac{1}{33}$ of a mile? (1 mile = 5280 feet)

3. In England, a person's weight is commonly given in stones. One English stone is equal to 14 pounds. If an English friend tells you he weighs thirteen stones, what is his weight in pounds?

4. The volume of a cement block is 7.25 in³. How many cubic centimeters is this? (1 cm = 0.393701 inches)

5. John is in an airport when 1 EUR will buy 1.39 USD. He has 215 euros to change. How many dollars does this get him? How many **quarters** could he buy for 215 euros?

For Exercises 6–11, suppose that an analog clock shows the time indicated. Find the smallest angle between the hour hand and the minute hand, in both degrees and radians.

6. 1 : 00
7. 3 : 00
8. 5 : 00
9. 6 : 00
10. 9 : 00
11. 10 : 00

For Exercises 12–19, convert the given angles to radians.

12. 90°
13. 45°
14. 450°
15. 150° 30′
16. 135°
17. 810°
18. 315°
19. 405° 40′ 15″

For Exercises 20–27, convert the given angles to degrees.

20. π rad
21. $\frac{\pi}{4}$ rad
22. $\frac{\pi}{6}$ rad
23. 3π rad
24. $\frac{11\pi}{6}$ rad
25. 5 rad
26. $\frac{8\pi}{3}$ rad
27. 6.5 rad

28. A **regular pentagon** is a geometric figure having five sides of equal length. Its interior angles will each measure 108°. Starting at a single point, use a ruler and a protractor to draw sides of length 5 cm and angles of 108° as accurately as possible, trying to arrive back at your starting point.

29. Use a ruler to draw a line segment A of length 6 cm. Then use a protractor to draw two angles at each endpoint measuring 58° and 47°, with A being the initial side. Use the protractor to find the measure of the third angle, and the ruler to estimate the lengths of the remaining sides as accurately as possible.

LESSON 3

The Six Trigonometric Ratios

Having discussed the concept of a real number and its role in measuring lengths and angles, we now turn to a geometric shape that involves both lengths and angles: the **triangle**. At its most basic level, a triangle is defined by three points or positions in space, which then give birth to three line segments connecting the points, therefore creating three angles.

Our main goal will be to determine all six of these values. We sometimes call this **solving a triangle**. That is, when given a few pieces of information about a triangle, we'll learn to use geometry and trigonometry to determine the remaining values.

We begin with a few facts and definitions related to triangles. It's certain that many of these ideas will be well known to our readers, such as the Pythagorean Theorem that was mentioned earlier in Lesson 1, or even the definition of a right angle itself. Nevertheless, we'll collect them here as a review and also for ease of reference. One idea that we want to particularly focus on is **similarity**, which is directly analogous to the definition of the radian measure of an angle. The following theorem is also absolutely essential, as it relates the three angles of a triangle.

Theorem: If a triangle is formed in a flat plane, then the sum of its interior angles is always $180°$ or π radians.

Lesson 3

Definition: Within a triangle, an angle is called **acute** if it measures less than 90°. An angle is called **right** if it measures exactly 90°, or **obtuse** if it measures between 90° and 180°.

According to the theorem and this definition, then, it is not possible for a triangle to contain two obtuse angles, or even two right angles. However, the measures of 90° and 180° seem to carry special significance, and this prompts the following definitions.

Definition: Two angles are called **complementary** if they add to 90°, and **supplementary** if they add to 180°.

We have just seen a number of adjectives that are used to describe different types of angles, and now we'll learn some that describe full triangles. The following definitions will be used very often.

Definition: A triangle is called **isosceles** if two of its sides are of the same length. Equivalently, a triangle is isosceles if two of its angles are congruent.

Definition: A triangle is called **equilateral** if all three of its sides are of the same length. Equivalently, a triangle is isosceles if all three of its angles are congruent.

Definition: A **right triangle** is a triangle having one right angle. The longest side, which has length c in the image below, is called the **hypotenuse**.

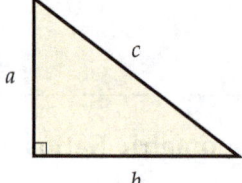

This definition leads us to one of the most important statements in the history of human thought.

The Pythagorean Theorem: Suppose that a triangle has three sides of lengths a, b, and c, with c being the largest of the three values. The triangle is a **right** triangle if, and only if, the lengths of the sides satisfy:

$$a^2 + b^2 = c^2.$$

One fact that is easily overlooked is that while the sides of right triangles satisfy the above relationship, they are the **only** triangles that do so. This means that, when coupled with the previous theorem about the interior angles, right triangles are particularly easy to work with.

LESSON 3

To further illustrate the importance of right triangles, note that **any** triangle at all can be split into two **right** triangles. We do this by introducing a new line segment from one vertex, intersecting the opposing side at a right angle. We sometimes refer to this extra line segment as an **altitude** of the triangle.

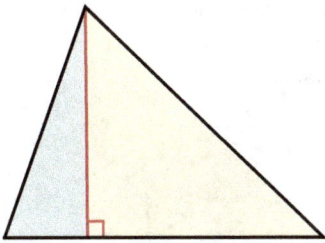

Obviously, every triangle has three distinct altitudes. These altitudes can be used in a large variety of ways, from bridging information between right triangles and other triangles to finding the **area** of a triangle via the formula

$$\text{Area} = \frac{1}{2}(b)(h).$$

Here h is the length of the altitude, and b is the length of the side intersecting it at a right angle.

The previous definitions have given us ways to describe individual triangles. These next few provide ways to **compare** two or more triangles in meaningful ways.

Recall that each side of a triangle has a length, described by a real number. Given two triangles, their corresponding sides may have lengths that are equal, in the sense that the real numbers describing them are equal. The same may be said of their angles and their measures. However, in the case that **all** of their angles and **all** of their side lengths are pairwise equal, the two triangles are still distinct geometric objects. In fact, there is no single real number that can completely describe a triangle, so that it might be "equal" to another triangle. Because of this, we use another word to describe this relationship.

Definition: Two triangles are called **congruent** if their corresponding sides are equal in length and their corresponding angles are equal in size.

This is certainly a powerful comparison, essentially the geometric equivalent of equality. It is so powerful, in fact, that for practical purposes it is often more powerful than we need. We would also like a way to describe two triangles that are the same **shape**, but not necessarily the same **size**.

LESSON 3

Definition: Two triangles are called **similar** if they have two equal angles, which then implies that **all** of their angles are equal.

As an example, the two triangles in the image at right have sides of length a, b, c and A, B, C, respectively. We can see that the angle formed by a and c is equal to the angle formed by A and C. Likewise, the angle between b and c is equal to the angle between B and C. Since the sum of the angles must be 180° in both cases, we can infer that the third angles must be equal as well.

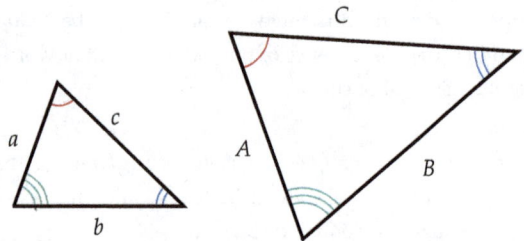

Another property of similar triangles, which is completely equivalent to the original definition, is that the ratios of their respective sides are equal:

$$\frac{A}{a} = \frac{B}{b} = \frac{C}{c}.$$

Note that each of these equalities can be algebraically rewritten in many ways, for example:

$$\frac{B}{A} = \frac{b}{a}, \quad \frac{A}{C} = \frac{a}{c}, \quad \frac{C}{B} = \frac{c}{b}.$$

If we recall the definition of radian measure, we'll remember that every angle defines a unique ratio between the length of an arc and the radius that defines it. This uniqueness allowed us to define the measure of an angle to be precisely this ratio.

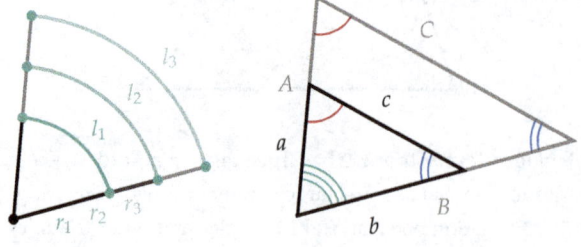

Here, if we superimpose one of our triangles onto another, similar one, we find that we again see the same equality of ratios. The difference is that here the sides a and b do not necessarily have the same length, resulting in several possible ratios for the triangle.

LESSON 3

Example: The triangles shown below have one right angle each, and one of their remaining angles is $\theta = 30°$. Therefore their third angles are also congruent, and the triangles are similar.

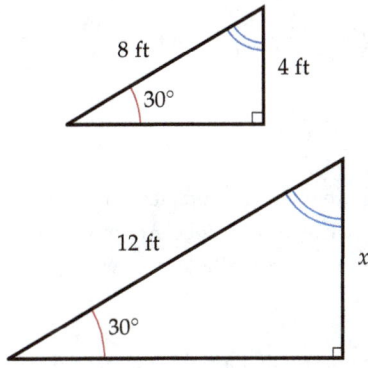

Since the ratios of their respective sides are equal, we can set them equal to each other:

$$\frac{4 \text{ ft}}{8 \text{ ft}} = \frac{x}{12 \text{ ft}}.$$

We simplify this and solve for x:

$$x = (12 \text{ ft})\left(\frac{4 \text{ ft}}{8 \text{ ft}}\right) = 6 \text{ ft}.$$

We've just found the length of x, the vertical side, which lies **opposite** to the angle 30°. The common ratio that we used to find x was of this opposite side to the hypotenuse:

$$\frac{\text{opposite}}{\text{hypotenuse}} = \frac{1}{2}.$$

Any similar triangle, regardless of size, will again have one angle of 30°, and the ratio between the side opposite to this angle and the hypotenuse will continue to equal 0.5.

Similarity is a property defined for all types of triangles, but for right triangles we find that knowing just one of its acute angles θ is enough to completely define its "similarity class." We still won't know anything about the triangle's **size**, but we will know everything about its **shape**. Moreover, this ratio that we've just found is fixed by this shape, and thus determined by θ.

However, this is just one of the possible ratios that we could have found for this particular triangle. We could have alternatively used the ratio

$$\frac{\text{hypotenuse}}{\text{opposite}} = \frac{2}{1} = 2,$$

which would have led us to the same solution. This ratio is also completely determined by θ.

27

LESSON 3

With three sides to the triangle, we find that the angle θ fixes not one but **six** different ratios. These ratios are so fundamental that we give each of them a name.

Definition: Suppose that a right triangle has one acute angle θ. One side of θ is the hypotenuse of the triangle, and the call the other side of θ the **adjacent** side of the triangle. The remaining side is **opposite** to θ.

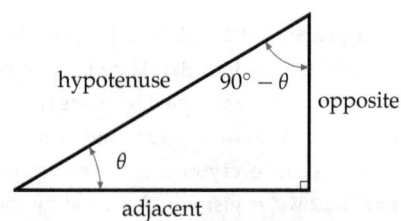

Then the following six ratios are fixed by θ:

$$\sin\theta = \frac{\text{opposite}}{\text{hypotenuse}} \qquad \csc\theta = \frac{\text{hypotenuse}}{\text{opposite}}$$

$$\cos\theta = \frac{\text{adjacent}}{\text{hypotenuse}} \qquad \sec\theta = \frac{\text{hypotenuse}}{\text{adjacent}}$$

$$\tan\theta = \frac{\text{opposite}}{\text{adjacent}} \qquad \cot\theta = \frac{\text{adjacent}}{\text{opposite}}$$

When referring to these ratios in writing, we use their full names; **sine, cosine, secant, cosecant, tangent,** and **cotangent**. In this usage the angle θ is not specified.

Directly from the definitions, the following should be clear:

$$\sec\theta = \frac{1}{\cos\theta} \qquad \tan\theta = \frac{\sin\theta}{\cos\theta} = \frac{1}{\cot\theta}$$

$$\csc\theta = \frac{1}{\sin\theta} \qquad \cot\theta = \frac{\cos\theta}{\sin\theta} = \frac{1}{\tan\theta}$$

These ratios are defined completely in terms of the sides of a triangle. If two of the sides are known, then it is usually unnecessary to actually know the angle θ, as the next example shows.

Example: Suppose that a right triangle has an acute angle θ, the adjacent side of the triangle has length 3 feet, and the opposite side has length 27 inches. Find $\cos\theta$.

We recall that the cosine of θ is the ratio of the adjacent length to the length of the hypotenuse. Since we are not given this second length, we will need to find it ourselves. Drawing a picture of our triangle, we quickly notice that the units of the two given lengths are not same, so we convert them to a common unit.

LESSON 3

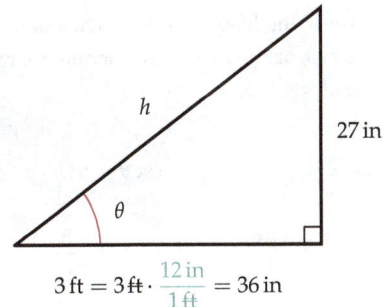

$$3\,\text{ft} = 3\,\text{ft} \cdot \frac{12\,\text{in}}{1\,\text{ft}} = 36\,\text{in}$$

The Pythagorean Theorem tells us that:

$$h^2 = (36\,\text{in})^2 + (27\,\text{in})^2$$
$$= (1296 + 729)\,\text{in}^2$$
$$= 2025\,\text{in}^2,$$

and therefore,

$$h = \sqrt{2025\,\text{in}^2} = 45\,\text{in}.$$

We can now find the cosine of θ from its definition. Note that in the ratio, the units cancel, leaving a pure ratio.

$$\cos\theta = \frac{\text{adjacent}}{\text{hypotenuse}} = \frac{36\,\text{in}}{45\,\text{in}} = \frac{4}{5}.$$

We have already seen that these trigonometric ratios have a lot in common with the angles that define them. While an angle is a ratio of an arc length to a linear length, it gives rise to a trigonometric ratio, which is a ratio of two linear lengths. Both are thus **dimensionless quantities**.

In fact, consider a triangle with a "small" angle θ, like the one below with a hypotenuse of length 5 and an opposite side of length 1.

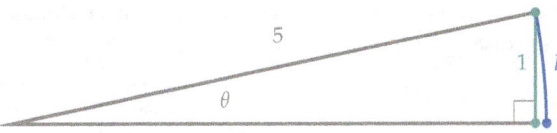

If we use the hypotenuse as a radius, the length of arc l that's created is very close to the length of the opposite side. Therefore the ratios θ and $\sin\theta$ are approximately equal:

$$\sin\theta = \frac{1}{5} \approx \frac{l}{5} = \theta.$$

However, this will not be true for larger angles, and the same relationship does not hold for the other trigonometric ratios. Let's do the work of finding out what those ratios are for this particular triangle.

29

LESSON 3

Example: Suppose that θ is an angle with
$$\sin\theta = \frac{1}{5}.$$

Find all five of the other trigonometric values of θ.

We again draw a triangle that represents this scenario, this time including the adjacent side a. We are given that $\sin\theta = 1/5$, but no units are associated with this problem. This is not an issue, because any units we might be using would cancel out when we form our new ratios, and be irrelevant in any case.

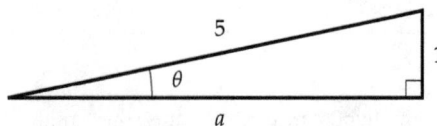

Knowing two sides of this right triangle, we use the Pythagorean Theorem to find the length of the third side, which is the side adjacent to θ:

$$a^2 + (1)^2 = (5)^2$$
$$a^2 = 25 - 1$$
$$a = \sqrt{24} = 2\sqrt{6}.$$

Now that we know the lengths of all three sides of the triangle, we can construct any of the trigonometric ratios that we like. We list all six here.

$$\sin\theta = \frac{1}{5} \qquad \csc\theta = 5$$

$$\cos\theta = \frac{2\sqrt{6}}{5} \qquad \sec\theta = \frac{5}{2\sqrt{6}}$$

$$\tan\theta = \frac{1}{2\sqrt{6}} \qquad \cot\theta = 2\sqrt{6}$$

One important difference between angles, and the six trigonometric ratios they create, is that angles carry their own units. Despite the fact that both are dimensionless quantities, we refer to angles in terms of either degrees or radians. Trigonometric ratios carry no such units. They are pure numbers that signify a relationship between two measurable quantities.

Thus far, we've been given the lengths of two sides of a right triangle, and we've learned more about the triangle based on that information. In the future, we may instead have information about an **angle** along with just one side. This is still enough information to determine all aspects of a triangle, though it may not seem so at first glance.

LESSON 3

Example: Suppose that a triangle has an acute angle of $\theta = 37°$, and the length of its adjacent side is 5.16 yards. Find the length of the hypotenuse.

With a quick picture to start, we see that we need a relationship between the adjacent side of the triangle and its hypotenuse.

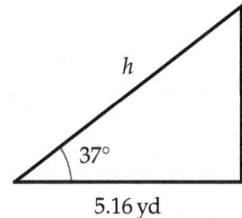

Looking at the definitions of the trigonometric ratios, we see that either cosine or secant will do. Using cosine, this tells us that
$$\cos(37°) = \frac{5.16\,\text{yd}}{h}.$$

If we knew the value of $\cos 37°$, the rest of this calculation would be straightforward. But how to find this ratio without a triangle to start from?

Fortunately, mathematicians, engineers, and scientists have been compiling tables of these trigonometric ratios for thousands of years, dating at least back to Hipparchus of Nicaea around 180–125 BCE. After constructing a right triangle of a given angle, through various methods, the ratios were carefully measured and recorded for future reference. The use of trigonometric tables continued well into the 20th century, and even today it is common to find such tables gracing the backs of relatively modern science textbooks.

In the present day we have much more portable devices for storing these types of data. Even a small, inexpensive calculator will tell us that
$$\cos(37°) \approx 0.7986.$$

This is an estimate, of course, because the actual value is irrational. But it will be sufficient for the problem at hand:
$$\cos(37°) = \frac{5.16\,\text{yd}}{h}$$
$$h = \frac{1}{\cos(37°)}(5.16\,\text{yd})$$
$$h \approx \frac{1}{0.7986}(5.16\,\text{yd})$$
$$h \approx 6.461\,\text{yd}.$$

It is now straightforward to find both the remaining angle and the length of the third side, solving this triangle.

LESSON 3

THE COMPLEMENTARY ANGLE THEOREM

We'll wrap up this lesson with a couple of helpful theorems. We saw earlier that some trigonometric ratios are the reciprocals of others, such as the cosine and secant ratios. However, the similarity of the names "sine" and "cosine" implies that there should be a relationship between them as well. In fact, there is a close relationship between each of the trigonometric ratios and its "co-ratio."

In the image to the right, we find a standard right triangle with one acute angle labeled θ. The side opposite to θ is labeled with length b, and the adjacent side is labeled with a. There is also the third, complementary angle, which has a measure of $90° - \theta$.

We notice that while the side labeled b is opposite to the angle θ, it is also adjacent to the angle $90° - \theta$. Since they share the same hypotenuse, we can surmise that

$$\sin\theta = \frac{b}{c} = \cos(90° - \theta).$$

The following relationships follow from the same argument. In each case, the value of a trigonometric ratio at an angle is equal to the value of its co-ratio at the corresponding complementary angle.

Theorem: If θ is an acute angle, then:

$$\sin\theta = \frac{b}{c} = \cos(90° - \theta) \qquad \cos\theta = \frac{a}{c} = \sin(90° - \theta)$$

$$\csc\theta = \frac{c}{b} = \sec(90° - \theta) \qquad \sec\theta = \frac{c}{a} = \csc(90° - \theta)$$

$$\tan\theta = \frac{b}{a} = \cot(90° - \theta) \qquad \cot\theta = \frac{a}{b} = \tan(90° - \theta)$$

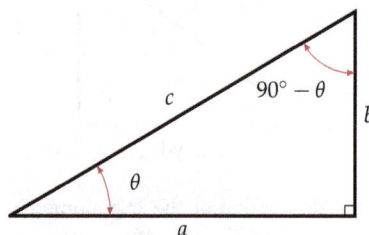

As a side note, we point out that when $\theta = 45°$, we have $\theta = 90° - \theta$. Thus the triangle in question is **isosceles**.

Since the corresponding side lengths of an isosceles triangle are also equal, we should expect that $\sin(45°) = \cos(45°)$, and similarly for the other cofunction pairs. We will explore this idea more in the next lesson.

LESSON 3

THE PYTHAGOREAN IDENTITIES

The next identities make use of the fact that the trigonometric ratios are defined using a right triangle, and therefore we may apply the Pythagorean Theorem to the lengths of its sides. We base the following statements on the angle θ shown in the image to the left, but we could have equally based them on either of the acute angles.

From the Pythagorean Theorem, the lengths of the three sides of the triangle satisfy

$$(\text{adjacent})^2 + (\text{opposite})^2 = (\text{hypotenuse})^2.$$

We write this in terms of the values a, b, and c, and then divide the equation by c^2. This leads directly into the definitions of the sine and cosine ratios.

$$(a)^2 + (b)^2 = (c)^2$$

$$\frac{a^2}{c^2} + \frac{b^2}{c^2} = 1$$

$$\left(\frac{a}{c}\right)^2 + \left(\frac{b}{c}\right)^2 = 1$$

$$(\sin\theta)^2 + (\cos\theta)^2 = 1.$$

As a matter of notation, we point out that the expression $\sin\theta^2$ is quite ambiguous, as it is not clear if one means to write $\sin(\theta^2)$ or $(\sin\theta)^2$. To help with this, and to save on parentheses, it is customary to write

$$\sin^2\theta = (\sin\theta)^2.$$

While the expression $\sin\theta^2$ should be avoided, is usually taken to mean $\sin(\theta^2)$ when it appears.

Now, from the Pythagorean Identity that we have just seen, we can obtain two more identities by dividing each side by either $\cos^2\theta$ and $\sin^2\theta$, respectively.

$$\sin^2\theta + \cos^2\theta = 1 \qquad\qquad \sin^2\theta + \cos^2\theta = 1$$

$$\frac{\sin^2\theta}{\cos^2\theta} + \frac{\cos^2\theta}{\cos^2\theta} = \frac{1}{\cos^2\theta} \qquad \frac{\sin^2\theta}{\sin^2\theta} + \frac{\cos^2\theta}{\sin^2\theta} = \frac{1}{\sin^2\theta}$$

$$\tan^2\theta + 1 = \sec^2\theta \qquad\qquad 1 + \cot^2\theta = \csc^2\theta$$

Theorem: If θ is an acute angle, then:

$$\sin^2\theta + \cos^2\theta = 1$$

$$\tan^2\theta + 1 = \sec^2\theta$$

$$1 + \cot^2\theta = \csc^2\theta$$

LESSON 3

EXERCISES

1. An isosceles triangle has two sides of length 5 cm and one side of length 3 cm. What is its area?

2. In the triangle below, what is $4x°$?

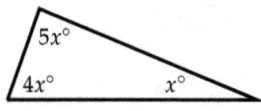

3. Is the above triangle a right triangle? How do you know?

4. The perimeters of two similar triangles are 14.7 cm and 51 cm, respectively. An altitude of the smaller triangle is 4 cm long. Find the length of the corresponding altitude of the larger triangle.

5. Suppose that a right triangle has one angle of $\theta = 52°$, and the length of the side opposite to θ is 2.58 meters. Find the length of the hypotenuse.

6. Suppose that a right triangle has one angle of $\theta = 43°$, and the length of the side adjacent to θ is 4.69 feet. Find the length of the opposite side.

7. Suppose that θ is an angle with $\sin\theta = 1/3$. Find all five of the other trigonometric values of θ.

8. Suppose that θ is an angle with $\cos\theta = 1/7$. Find all five of the other trigonometric values of θ.

Given that $\tan\theta = 2$, use trigonometric identities to find:

9. $\sec^2\theta$ 10. $\cot\theta$ 11. $\cot(90° - \theta)$ 12. $\csc^2\theta$

Given that $\sec\theta = 3$, use trigonometric identities to find:

13. $\cos\theta$ 14. $\tan^2\theta$ 15. $\csc(90° - \theta)$ 16. $\sin^2\theta$

17. A 24-foot extension ladder is rested against a wall, making an angle of 72° with the flat ground. How far up the wall is the top of the ladder? Round your answer to two decimal places.

18. A flat-screen television has an aspect ratio of 16 : 9, which signifies the ratio of the lengths of its sides. If we buy a television with a 47-inch diagonal, how long are each of its sides? Round your answer to two decimal places.

LESSON 4

Standard Triangles

Now that we've defined the six trigonometric ratios for a general right triangle, we'll turn to a few specific examples.

The first triangles to examine are commonly used by architects and draftsmen when making blueprints or technical drawings. Sometimes called **set squares**, one has two acute angles of 45°, and the other has acute angles of 30° and 60°. Combining several of these triangles also allows us to quickly and accurately draw other angles like 15° or 75°. Because they are so commonly used, we memorize their trigonometric ratios rather than resort to a calculator or computer every time we need them.

Memorizing these ratios has an added advantage; we'll be using their **exact** values in terms of square roots. This is not generally the case when resorting to calculators, which give decimal approximations of the irrational numbers.

Afterward, we'll look at triangles with a different practical purpose. Rather than being used in drawing or planning stages, these triangles are used in an actual construction setting to determine if two walls or beams meet at a right angle. This same method has been used for nearly all of written history, and even today it remains the simplest and most accurate way of making such a verification.

LESSON 4

THE 45/45/90 TRIANGLE

For our first example, we begin with the acute angle $\theta = 45°$. Drawing a right triangle with this angle, we label the adjacent and opposite sides as a and b, respectively.

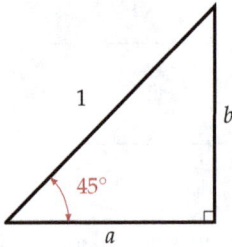

We immediately find that the third, complementary angle is also 45°. As we noted in the last lesson, our right triangle is **isosceles**, and $a = b$.

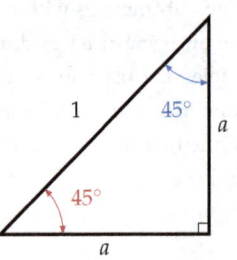

We can now use the Pythagorean Theorem to find the length of each side of our triangle.

$$a^2 + b^2 = 1^2$$
$$2a^2 = 1$$
$$a^2 = \frac{1}{2}$$
$$a = b = \frac{\sqrt{2}}{2}.$$

Using our definitions of the trigonometric functions, we conclude the following theorem.

Theorem: The following identities hold for $\theta = 45°$

$$\sin\left(\frac{\pi}{4}\right) = \frac{b}{1} = \frac{\sqrt{2}}{2} \qquad \cos\left(\frac{\pi}{4}\right) = \frac{a}{1} = \frac{\sqrt{2}}{2}$$

$$\csc\left(\frac{\pi}{4}\right) = \frac{1}{b} = \sqrt{2} \qquad \sec\left(\frac{\pi}{4}\right) = \frac{1}{a} = \sqrt{2}$$

$$\tan\left(\frac{\pi}{4}\right) = \frac{b}{a} = 1 \qquad \cot\left(\frac{\pi}{4}\right) = \frac{a}{b} = 1$$

Note that since the triangle is isosceles, by the Complementary Angle Theorem we should expect the value of each ratio to be equal to that of its co-ratio, and that is indeed the case.

LESSON 4

Example: Two men are standing on the rooftops of two buildings that are the same height. The men are 100 feet from each other, and each of them is holding one end of the same rope. The rope is supporting a heavy object, so it is pulled down at an angle of 45° from the horizontal on both sides. How long is the rope? How far below the rooflines is the object?

Since the men are both standing at the same height, and the angle between the rope and the horizontal is the same on each side, this system forms an isosceles triangle that is shown below.

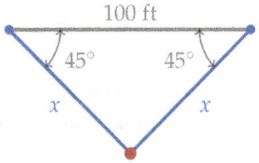

More specifically, this is a 45/45/90 triangle with a hypotenuse measuring 100 feet. We have just learned that

$$\sin 45° = \cos 45° = \frac{\sqrt{2}}{2}.$$

It does not particularly matter which of these we use.

If we focus on the left-most angle and left-most x, then the definition of cosine tells us that

$$\cos 45° = \frac{\text{adjacent}}{\text{hypotenuse}}$$

$$\frac{\sqrt{2}}{2} = \frac{x}{100 \text{ ft}}$$

$$x = \frac{\sqrt{2}}{2}(100 \text{ ft})$$

$$x = 50\sqrt{2} \text{ ft.}$$

Since x represents only half of the total length of the rope, we conclude that the rope is $100\sqrt{2}$ feet long. (For reference, this is approximately 141.42 feet.) For the second question, note that the vertical height of the object forms two smaller 45/45/90 triangles:

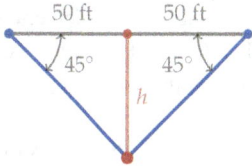

Each of these smaller triangles is again isosceles, so the height h is half of the distance between the two men, or $h = 50$ ft.

37

LESSON 4

THE 30/60/90 TRIANGLE

For our next standard triangle, we will begin with the acute angle $\theta = 60°$. Again, we label the adjacent and opposite sides as a and b, respectively.

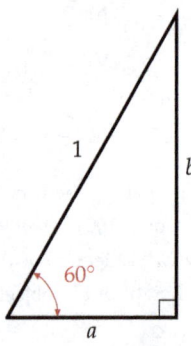

We find that the remaining complementary angle is 30°. This triangle is therefore not isosceles, so we must try a different approach from the last example.

Instead, we will draw a second, identical triangle adjacent to the one that we have. The larger of the duplicate angles is again a 60° angle, and sum of the two 30° angles at the top of the image is a third. Therefore two triangles together create a single **equilateral** triangle.

All sides of this equilateral triangle have the same length as the original hypotenuse, which is 1 unit. Since a is half of that length across the triangle's base, we find that $a = 0.5$:

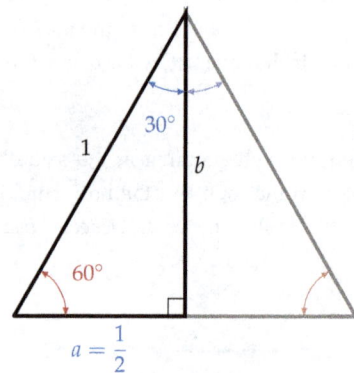

Now that we know two side lengths, we can use the Pythagorean Theorem to find the remaining one, b.

$$\left(\frac{1}{2}\right)^2 + b^2 = 1^2$$
$$b^2 = 1 - \frac{1}{4}$$
$$b^2 = \frac{3}{4}$$
$$b = \frac{\sqrt{3}}{2}.$$

We have now obtained the following trigonometric values for $\theta = 60°$ and $\theta = 30°$. Note that these tables are an excellent illustration of the Complementary Angle Theorem that we learned in Lesson 3.

Theorem: The following identities hold for $\theta = 60°$

$$\sin\left(\frac{\pi}{3}\right) = \frac{b}{1} = \frac{\sqrt{3}}{2} \qquad \cos\left(\frac{\pi}{3}\right) = \frac{a}{1} = \frac{1}{2}$$

$$\csc\left(\frac{\pi}{3}\right) = \frac{1}{b} = \frac{2\sqrt{3}}{3} \qquad \sec\left(\frac{\pi}{3}\right) = \frac{a}{1} = 2$$

$$\tan\left(\frac{\pi}{3}\right) = \frac{b}{a} = \sqrt{3} \qquad \cot\left(\frac{\pi}{3}\right) = \frac{a}{b} = \frac{\sqrt{3}}{3}$$

Theorem: The following identities hold for $\theta = 30°$

$$\sin\left(\frac{\pi}{6}\right) = \frac{b}{1} = \frac{1}{2} \qquad \cos\left(\frac{\pi}{6}\right) = \frac{a}{1} = \frac{\sqrt{3}}{2}$$

$$\csc\left(\frac{\pi}{6}\right) = \frac{1}{b} = 2 \qquad \sec\left(\frac{\pi}{6}\right) = \frac{a}{1} = \frac{2\sqrt{3}}{3}$$

$$\tan\left(\frac{\pi}{6}\right) = \frac{b}{a} = \frac{\sqrt{3}}{3} \qquad \cot\left(\frac{\pi}{6}\right) = \frac{a}{b} = \sqrt{3}$$

Example: Suppose that an equilateral triangle has sides of length 1 m each, and a circle is superscribed so that is passes through all three vertices. Find the radius of the circle.

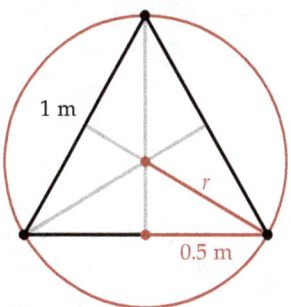

Introducing an **altitude** in an equilateral triangle creates two 30/60/90 triangles, as we've just seen. But if we introduce all **three** altitudes, this creates **six** smaller 30/60/90 triangles. Each longer side measures 0.5 m, and each hypotenuse is a radius; both are marked in red above. Since the angle between them is 30°:

$$\cos 30° = \frac{\sqrt{3}}{2} = \frac{0.5\,\text{m}}{r}.$$

Solving the equality on the right gives us $r = \frac{1}{\sqrt{3}}$ m.

LESSON 4

While it may be tempting to simply memorize the values in the previous theorems (which you should), it is far more useful to remember the following pictures of the Standard Triangles. Remember that these triangles can be drawn in any necessary orientation.

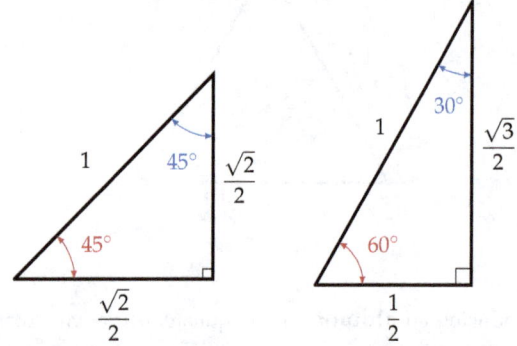

Any trigonometric values needed for these standard angles can be obtained from these pictures. Again, despite the fact that half of these side lengths are irrational, we still have exact expressions for them.

We will refer to these triangles extensively in future examples, both here and in later lessons. Notice that in each of the following questions, we are asked to find the **exact** value of an expression.

This is a signal that we should not be using a calculator, but instead use our knowledge of the standard angles along with the rules of arithmetic.

Example: Find the exact value of $\sec\left(\frac{\pi}{4}\right) + \sin\left(\frac{\pi}{6}\right)$.

For the first expression of this sum, we look to the 45/45/90 triangle from the previous page. Secant is defined as the ratio of the hypotenuse to the adjacent side. We find these sides' lengths on the triangle, and divide:

$$\sec\left(\frac{\pi}{4}\right) = \frac{1}{\sqrt{2}/2} = \frac{2}{\sqrt{2}} = \sqrt{2}.$$

For the second expression, we look to the 30/60/90 triangle. There we find the lengths of the opposite side and hypotenuse, and divide appropriately:

$$\sin\left(\frac{\pi}{6}\right) = \frac{1/2}{1} = \frac{1}{2}.$$

To add these two values together, we find a common denominator, and obtain

$$\sec\left(\frac{\pi}{4}\right) + \sin\left(\frac{\pi}{6}\right) = \sqrt{2} + \frac{1}{2} = \frac{2\sqrt{2}}{2} + \frac{1}{2} = \frac{2\sqrt{2}+1}{2}.$$

LESSON 4

Example: Find the exact value of $\cot^3(30°)\csc(60°)$.

Both angles in this product belong to the 30/60/90 triangle, so we refer to that triangle throughout our argument. In the first expression, cotangent is defined as the ratio of the adjacent side to the opposite side, so we divide those two lengths accordingly:

$$\cot(30°) = \frac{\sqrt{3}/2}{1/2} = \sqrt{3}.$$

In the example, this expression is cubed, so we use the laws of exponents to write:

$$\cot^3(30°) = \left(\sqrt{3}\right)^3 = \left(3^{1/2}\right)^3 = 3^{\frac{3}{2}}.$$

Next we turn to the second factor. Cosecant is the ratio of the opposite side to the hypotenuse, so we divide these:

$$\csc(60°) = \frac{1}{\sqrt{3}/2} = \frac{2\sqrt{3}}{3}.$$

Now we multiply our two values together, using the rules of multiplication and exponents to simplify:

$$\cot^3(30°)\csc(60°) = \left(3^{\frac{3}{2}}\right)\left(\frac{2\sqrt{3}}{3}\right) = 2\left(3^{\frac{3}{2}+\frac{1}{2}-1}\right) = 6.$$

When working with the exact values from these two triangles, we rely heavily on the rules of arithmetic, particularly those of square roots and exponents. These rules allow us to combine our values in a meaningful way while maintaining the exactness of our solution.

Suppose that we had instead estimated the values of $\cot 30°$ and $\csc 60°$ in the last problem, settling on something like

$$\cot 30° \approx 1.732,$$

$$\csc 60° \approx 1.1547.$$

Blind calculation with a calculator would then give us a decimal value very close to 6. But it would be difficult or impossible for us to know if the actual value should be **equal** to 6, rather than some irrational value very close to it.

Our final class of triangles also involves this concept of exactness. We already know that the side lengths a, b, and c of a right triangle must satisfy the Pythagorean Theorem. So even if two of the sides have integer lengths, by virtue of the theorem, the third side will **almost always** be irrational. However, in certain very rare situations, all three sides will have rational or even **integer** lengths.

LESSON 4

PYTHAGOREAN TRIPLES

If we are concerned only with the three side lengths of a triangle, then we can express our desired relationship in a completely algebraic way, without involving the angles at all.

Definition: A **Pythagorean triple** is a set of three **integers** $a, b, c \in \mathbb{Z}$ satisfying the equation $a^2 + b^2 = c^2$.

For example, the numbers 1, 2, 3 are all integers, but do not form a Pythagorean triple because

$$1^2 + 2^2 = 1 + 4 = 5 \neq 3^2.$$

If we modify this slightly by taking the triple $\left(1, \sqrt{3}, 2\right)$, we find that these numbers do satisfy the equation:

$$1^2 + \left(\sqrt{3}\right)^2 = 1 + 3 = 4 = 2^2.$$

We also note that these three lengths form a 30/60/90 triangle. However, this triple is not Pythagorean, since $\sqrt{3} \notin \mathbb{Z}$.

The smallest example of three such numbers is $(3, 4, 5)$:

$$3^2 + 4^2 = 9 + 16 = 25 = 5^2.$$

The usefulness of this triple has been known for thousands of years. Imagine that we are constructing a building, and we have two beams that we would like to be **perpendicular**, meaning that they should meet at a right angle. To test this, we measure from the corner 3 units along one beam, and 4 units along the other, marking each location.

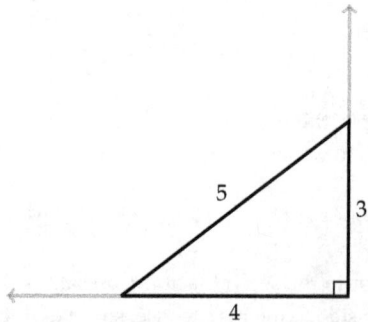

Then, we measure diagonally between the marked locations. If the diagonal length measures slightly less than 5 units, then the angle between the beams is slightly acute. If it is slightly more than 5 units, the angle is slightly obtuse. But if the diagonal length measures exactly 5 units, then we have a Pythagorean triple, and the three lengths form a right triangle.

LESSON 4

Here we are using the fact that **if** the lengths a, b, and c of a triangle satisfy the equation $a^2 + b^2 = c^2$, **then** the angle opposite c is a right angle. In other words, not only do the side lengths of right triangles satisfy this equation, but they are also the **only** triangles with this property.

We can use any units of measurement that we like for this type of verification, though using larger units like feet or meters gives us a greater degree of precision than do smaller units like inches or centimeters. In fact, if space allows, we can gain precision with larger Pythagorean triples like $(6, 8, 10)$ or $(5, 12, 13)$:

$$6^2 + 8^2 = 36 + 64 = 100 = 10^2,$$

$$5^2 + 12^2 = 25 + 144 = 169 = 13^2.$$

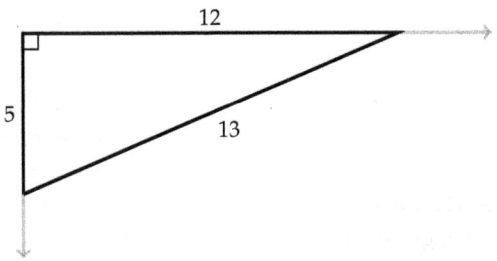

Despite their rarity, there are still infinitely many Pythagorean triples. One method of generating such triples, attributed to Euclid of Alexandria around 300 BCE, begins with any two integers m and n. Then we construct:

$$a = m^2 - n^2,$$
$$b = 2mn,$$
$$c = m^2 + n^2.$$

Showing that these satisfy our equation is straightforward:

$$\begin{aligned} a^2 + b^2 &= (m^2 - n^2)^2 + (2mn)^2 \\ &= m^4 - 2m^2n^2 + n^4 + 4m^2n^2 \\ &= m^4 + 2m^2n^2 + n^4 \\ &= (m^2 + n^2)^2 \\ &= c^2. \end{aligned}$$

What is more, once we have a Pythagorean triple (a, b, c), a new one can be generated by multiplying each number in the triple by a common constant. For example, the Pythagorean triple $(6, 8, 10)$ is generated by multiplying $(3, 4, 5)$ by a factor of 2. The triple $(5, 12, 13)$ cannot be generated in this way, however.

LESSON 4

To describe this "independence" of Pythagorean triples, we use the following definition.

Definition: A Pythagorean triple is called **primitive** if there is no factor that is common to all three of the integers a, b, and c.

We may have noticed that our definitions for Pythagorean triples include only **integer** triples, and not rational triples. It is precisely the idea of multiplying by a common factor that allows this. Consider the following triple:

$$\left(\frac{8}{9}, \frac{7}{2}, \frac{65}{18}\right)$$

It is not obvious at first glance, but these rational numbers satisfy the equation $a^2 + b^2 = c^2$:

$$\left(\frac{8}{9}\right)^2 + \left(\frac{7}{2}\right)^2 = \frac{64}{81} + \frac{49}{4} = \frac{4225}{324} = \left(\frac{65}{18}\right)^2.$$

However, since they are all rational, we can multiply by an appropriate factor to turn all three into integers. Multiplying each by 18 gives us the triple $(16, 64, 65)$, and we can verify that they still satisfy the Pythagorean Theorem:

$$16^2 + 63^2 = 256 + 3969 = 4225 = 65^2.$$

In this way the so-called rational triples are always a constant multiple of a primitive Pythagorean triple, though we don't call them Pythagorean triples themselves. This may be because their usefulness in practical applications is questionable, compared to integer triples.

There are also infinitely many **primitive** Pythagorean triples. In fact, every integer greater than 2 is part of at least one Pythagorean triple, either primitive or non-primitive, depending on the integer. The sixteen lowest primitive Pythagorean triples are listed below.

(3, 4, 5)	(5, 12, 13)	(8, 15, 17)	(7, 24, 25)
(20, 21, 29)	(12, 35, 37)	(9, 40, 41)	(28, 45, 53)
(11, 60, 61)	(16, 63, 65)	(33, 56, 65)	(48, 55, 73)
(13, 84, 85)	(36, 77, 85)	(39, 80, 89)	(65, 72, 97)

The concession that we make for the simplicity of these triangles involves their angles. When measured in degrees, it is not possible for one of these triangles to have rational angles; right triangles with three rational sides **and** three rational angles do not exist. However, in upcoming lessons we will learn how to determine these angles, rational or not.

LESSON 4

EXERCISES

For Exercises 1–6, find the exact value of each expression, without using a calculator.

1. $\cos^2 30° - \sin^2 30°$

2. $\dfrac{\sin 30° + \cos 60°}{2}$

3. $\sec 60° + \cot 30°$

4. $8(\sin 120°)(\cos 120°)$

5. $\dfrac{4 \sin 300° + 2 \cos 30°}{3}$

6. $2 \csc \dfrac{\pi}{4} - \sec \dfrac{\pi}{3} \cos \dfrac{\pi}{6}$

7. Find the exact length of one side of a square with a diagonal that measures 16 centimeters.

8. Find the perimeter of a square that has a diagonal that measures 10 meters.

9. Find the exact length of the diagonal of a square with a 15 decimeter side.

10. The perimeter of a square is 48 meters. Find the length of a side and the length of a diagonal.

11. Suppose that a **regular hexagon** has sides that measure 10 meters each. Find the exact area of the hexagon.

12. A conveyor belt is used to take wheat from ground level to the top of a silo for storage. If the conveyor belt is 124 feet long and the silo is 62 feet high, how far is the end of the conveyor from the base of a silo?

Is a triangle with the given side lengths a **right** triangle?

13. 20, 99, 101   14. 24, 143, 147   15. 119, 120, 169

Determine if each of the Pythagorean triples is **primitive**.

16. $(48, 189, 195)$ 17. $(85, 132, 157)$ 18. $(88, 165, 187)$

19. Use a ruler to draw a right triangle with side lengths of 3 in, 4 in, and 5 in. Label the smaller of the two acute angles as α, and the larger as β.

20. Use a protractor to estimate α from Exercise 19. Then use a calculator to find the tangent of your estimated angle. Is this tangent larger or smaller than 0.75? Adjust your estimate accordingly, and try again. Working this way, find α to within three decimal places.

21. Repeat Exercise 20 to find the angle β to within three decimal places.

LESSON 5

The Unit Circle

Throughout the last two lessons, the trigonometric ratios have been defined by the acute angles of a right triangle. For the standard angles 30°, 45°, and 60°, you should have memorized each of the six trigonometric values. For any other acute angle, you can estimate their trigonometric values using a calculator.

However, at this stage, our definitions might prompt a few interesting questions:

1. Can an angle be **negative**? If so, how do we interpret any resulting triangles or trigonometric ratios?

2. What about **obtuse** angles? What is the sine of 125 degrees? How do we talk about $\sin\theta$ as the ratio of the "opposite over the hypotenuse" when there is no right triangle with such an angle? What about even larger angles that are greater than 360°?

3. Even if we can concoct some way to make sense of $\sin(125°)$, via some sort of reflection of triangle or a rule about positive and negative signs, what do we do about $\sin(90°)$? There **can't** be a triangle here to talk about, since the opposite side and the hypotenuse of such a "triangle" would be **parallel**.

LESSON 5

As we've mentioned before, the six trigonometric ratios of a triangle have been with us for several millennia. Ancient mathematicians compiled tables of these ratios to varying degrees of accuracy, for very fine measurements of angles between 0° and 90°. But only in the relatively recent past have the definitions been considered in the kind of general setting indicated by these questions.

It's clear that we're going to need a new definition of $\sin \theta$, one that's more flexible for different θ. We would very much like to end up with sine defined as a **function** with a well-defined input and output; it should be a rule that assigns some real number to a given angle of any measure. And of course, as long as we're defining the sine function, we'll define all of the trigonometric functions. For now, we can forget all of their previous definitions.

We'll also introduce a new **tool** to help establish the new definitions. The triangles and circles we've worked with so far were freely existing geometric objects, not having any particular position or orientation in space. In this context we can measure features of an object only with respect to other features on the **same object**. If we want to describe position or orientation, or compare one object to another, we must introduce an ambient **coordinate system**.

Every coordinate system consists of the following choices:

1. A choice of **origin**.

2. A standard unit of **distance**.

3. A choice of **orientation** for each dimension.

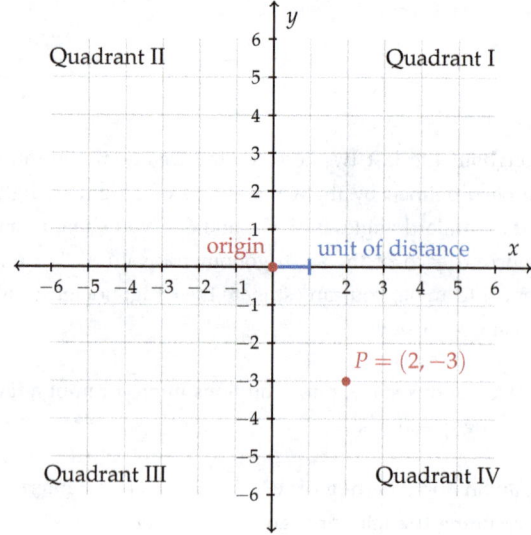

Note that the coordinate system in our image consists of two dimensions, spanned by the two **coordinate axes**, which intersect at the origin. We've marked each axis at regular intervals, the length of each interval indicating the standard unit of length for our coordinate system. We've also given orientation to each axis by choosing the positive and negative directions. Finally, the four quadrants of the coordinate system have been labeled for easier reference.

Each position or point P in our coordinate system can be identified by a unique ordered pair $P = (x, y)$, with x indicating the horizontal distance of the point from the y-axis, and y indicating the vertical distance of the point from the x-axis. We will sometimes refer to this collection of points as the x, y-**plane**.

Now that we have a coordinate system, we can redefine our concept of angle for this context. We do this in a way that is consistent with the conventions we set in Lesson 2.

Definition: An angle θ in the x, y-plane is always measured from the **positive x-axis**, which is then the initial side of θ. A positive angle is measured in the **counterclockwise** direction, ending at the terminal side of θ. We interpret a negative angle as being measured **clockwise** from the positive x-axis.

Our new coordinate system provides us with everything we need to redefine the sine and cosine functions. Into this system we place the **Unit Circle**, which is the set of all points that are distance one from the origin.

According to the Pythagorean Theorem, this is also the set of points (x, y) satisfying the equation $x^2 + y^2 = 1$.

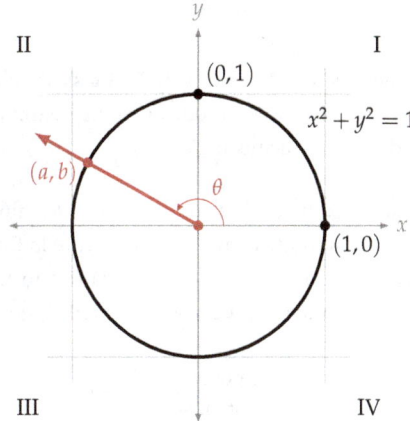

Now, choose any angle θ. Theta may be either greater or less than $360°$, positive, negative, or even zero. We have drawn the terminal side of such an angle in red in the figure above.

LESSON 5

This terminal side will intersect the Unit Circle at exactly one point (a,b), which is also drawn in red. That is, the values of *a* and *b* are uniquely determined by θ.

Definition: Let (a, b) be the unique point at which the terminal side of the angle θ intersects the Unit Circle. Then

$$\cos\theta = a,$$
$$\sin\theta = b.$$

Following these definitions, our very first task should be to verify that they coincide with our previous definitions for the sine and cosine of acute angles.

Therefore, suppose that θ is acute, so that its terminal side lies in the first quadrant, as shown in the figure to the right. The terminal side of θ creates a right triangle in the first quadrant, in which the hypotenuse has length 1, giving:

$$\sin\theta = \frac{\text{opposite}}{\text{hypotenuse}} = \frac{b}{1} = b,$$

$$\cos\theta = \frac{\text{adjacent}}{\text{hypotenuse}} = \frac{a}{1} = a.$$

Therefore this new definition agrees with our previous one.

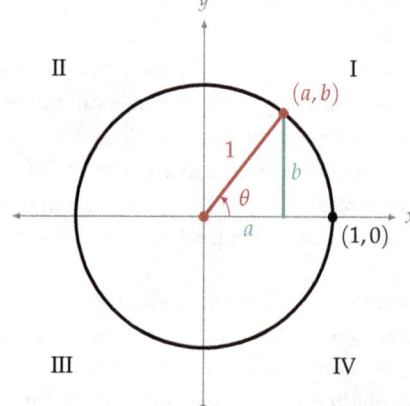

But in addition to providing us with a new definition, the Unit Circle is also a powerful tool for **evaluating** trigonometric functions. Specifically, the Unit Circle will allow us to keep track of the signs of trigonometric values, based on the quadrant in which the angles lie.

Example: Find both $\sin\left(\frac{2\pi}{3}\right)$ and $\cos\left(\frac{2\pi}{3}\right)$.

We begin by drawing the Unit Circle along with the ray that makes an angle of 120° with the positive *x*-axis. This identifies a point (a, b) on the Unit Circle.

To find its coordinates, we draw a vertical line down from (a,b) to the x-axis. This creates a right triangle with a hypotenuse of length 1.

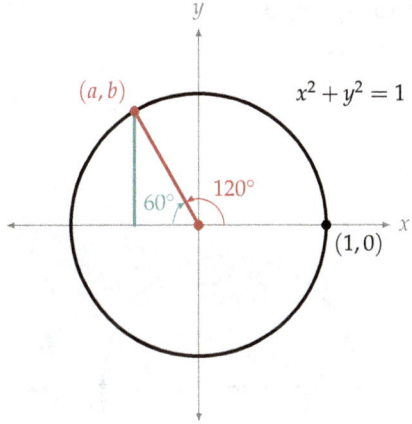

The angle that this triangle makes with the x-axis is

$$\theta = 180° - 120° = 60°.$$

We recognize this right triangle as a 30/60/90 triangle, which we studied in Lesson 4. This triangle is oriented so that it lies in the second quadrant, but for now we will just find the lengths of its sides.

From our previous work in Lesson 4:

$$\cos\left(\frac{\pi}{3}\right) = \frac{1}{2}, \quad \sin\left(\frac{\pi}{3}\right) = \frac{\sqrt{3}}{2}.$$

With a hypotenuse of length one, these tell us the distance of (a,b) from the y- and x-axes, respectively. In Quadrant II the coordinate a is negative, and the coordinate b is positive:

$$(a,b) = \left(-\frac{1}{2}, \frac{\sqrt{3}}{2}\right).$$

Therefore our trigonometric values are

$$\cos\left(\frac{2\pi}{3}\right) = -\frac{1}{2},$$

$$\sin\left(\frac{2\pi}{3}\right) = \frac{\sqrt{3}}{2}.$$

In this example, the original angle 120° led to an accompanying **acute** angle of 60°, and it was with this acute angle that we performed any calculations. This will be how we evaluate all such trigonometric functions, and therefore this acute angle warrants a title.

Definition: Let θ be an angle measured from the positive x-axis. The acute angle formed by the terminal side of θ and the x-axis is called the **reference angle** for θ.

LESSON 5

One important consequence of our new definitions is that we can deal with angles of any magnitude. For this we recall that in Lesson 2, two angles are called **coterminal** if they differ by a multiple of 360° or 2π radians. That is, two angles θ and ϕ are coterminal if they point in the same direction in the x,y-plane. In this situation the two angles have the **same reference angle**.

Theorem: Let θ and ϕ be coterminal angles. Then

$$\sin\theta = \sin\phi,$$
$$\cos\theta = \cos\phi.$$

Example: Find $\sin(-17\pi)$.

In this example we find that the angle θ, written in degrees, is not in the interval $[0°, 360°)$. We will find the coterminal angle by adding or subtracting $2\pi = 360°$ as many times as needed. For example, adding one full circle gives

$$-17\pi + 2\pi = -15\pi.$$

This is closer to the interval $[0°, 360°)$, but still not in it. There will only be one such angle. In this case, the only coterminal angle in this interval is $\phi = \pi$, as:

$$-17\pi + 9(2\pi) = \pi.$$

In this case, the reference angle is actually equal to zero, and we find that there is no "reference triangle" at all. But this is not a major issue, since sine is no longer defined in terms of a triangle, but rather as the y-coordinate of a point (a,b) in the x,y-plane. This point lies on the x-axis, so we conclude that

$$\sin(-17\pi) = \sin\pi = 0.$$

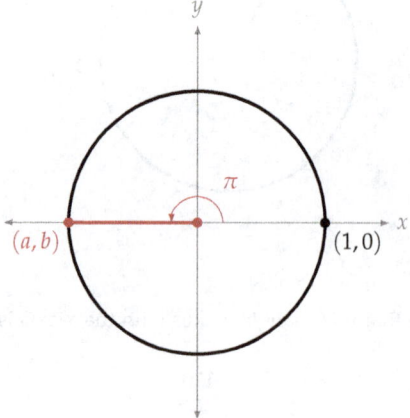

Of course, once we have these definitions of sine and cosine, we can define our remaining four trigonometric functions: tangent, cotangent, secant, and cosecant.

LESSON 5

Definitions: Suppose that θ is any angle that is not an integer multiple of $90°$. Then:

$$\sec\theta = \frac{1}{\cos\theta} \qquad \tan\theta = \frac{\sin\theta}{\cos\theta}$$

$$\csc\theta = \frac{1}{\sin\theta} \qquad \cot\theta = \frac{\cos\theta}{\sin\theta}$$

We will demonstrate these definitions with another example. In this example, we will use a nonstandard angle, meaning that we will need to estimate our trigonometric values using a calculator. However, the method will be identical to the previous example.

Example: Find $\cot(553°)$.

We immediately note that our angle is greater than $360°$, so we first find the coterminal angle in the interval $[0°, 360°)$. In this case, the coterminal angle is

$$\alpha = 553° - 360° = 193°.$$

The terminal side of this angle lies in the third quadrant, and we find the **reference** angle

$$\beta = 193° - 180° = 13°.$$

A calculator tells us that

$$\cos(13°) \approx 0.97437, \quad \sin(13°) \approx 0.22495.$$

The point (a, b) then has the approximate coordinates

$$(a, b) \approx (-0.97437, -0.22495).$$

Finally, we use the definition of cotangent as the ratio of cosine to sine, to arrive at our solution:

$$\cot(553°) = \frac{\cos(553°)}{\sin(553°)} \approx \frac{-0.97437}{-0.22495} \approx 4.3315.$$

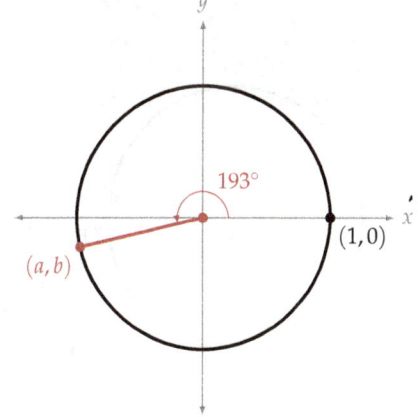

LESSON 5

It is very important to consider that these definitions exclude the angles θ that are multiples of $90°$. This is because these remaining four functions are defined as **ratios** of sine and cosine.

While sine and cosine are defined for any angle, these four will not be defined for angles at which their denominators are equal to zero. In these cases we simply say that their trigonometric values are **undefined**.

Example: Find both $\sec\left(-\frac{17\pi}{2}\right)$ and $\csc\left(-\frac{17\pi}{2}\right)$.

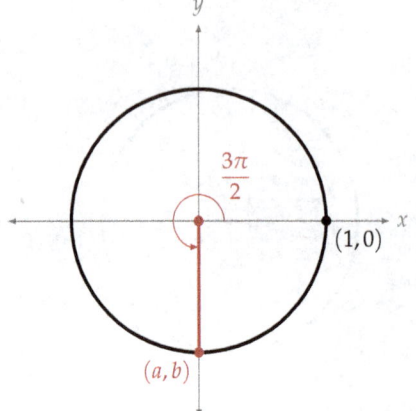

Here the angle is given in radians, so to find the coterminal angle, we will add multiples of 2π radians until we find an angle in the range $[0, 2\pi)$. We find that

$$-\frac{17\pi}{2} + 5(2\pi) = -\frac{17\pi}{2} + \frac{20\pi}{2} = \frac{3\pi}{2},$$

so the angle $\theta = -\frac{17\pi}{2}$ is coterminal to $\phi = 270°$.

Note that our reference angle in this case is again not acute, but instead the point (a, b) lies directly on one of the coordinate axes. More precisely, it has coordinates $(a, b) = (0, -1)$. We find

$$\cos\left(-\frac{17\pi}{2}\right) = \cos(270°) = 0,$$

$$\sin\left(-\frac{17\pi}{2}\right) = \sin(270°) = -1.$$

Therefore, as cosine is the reciprocal of sine, we have

$$\csc\left(-\frac{17\pi}{2}\right) = \frac{1}{\sin(270°)} = -1.$$

However, since $\cos(270°) = 0$, the reciprocal defined by secant is **undefined**:

$$\sec\left(-\frac{17\pi}{2}\right) = \frac{1}{\cos(270°)}.$$

For the remainder of this lesson we'll look at some of the properties of trigonometric functions we saw previously. We can also establish a few new properties that are accessible from the new definitions.

THE PYTHAGOREAN IDENTITIES

For any angle θ, the corresponding point on the Unit Circle

$$(a, b) = (\cos\theta, \sin\theta)$$

satisfies the equation $x^2 + y^2 = 1$, by definition. Therefore the Pythagorean Identity $\sin^2\theta + \cos^2\theta = 1$ follows immediately from the new definitions of sine and cosine, for any angle θ, even for angles like $0°$ or $90°$ when a corresponding right triangle does not exist.

Starting from that identity, we can obtain the two remaining Pythagorean Identities just as we did in Lesson 3, assuming that the trigonometric values are defined in the first place.

Theorem: If θ is any angle, then:

$$\sin^2\theta + \cos^2\theta = 1$$

$$\tan^2\theta + 1 = \sec^2\theta$$

$$1 + \cot^2\theta = \csc^2\theta$$

BOUNDEDNESS OF SINE AND COSINE

The values $\cos\theta$ and $\sin\theta$ denote the x- and y-coordinates of a point on the Unit Circle, which is defined by the equation $x^2 + y^2 = 1$. It is clear in the image that the y-coordinate can never be greater than 1 nor less than -1, and likewise for the x-coordinate.

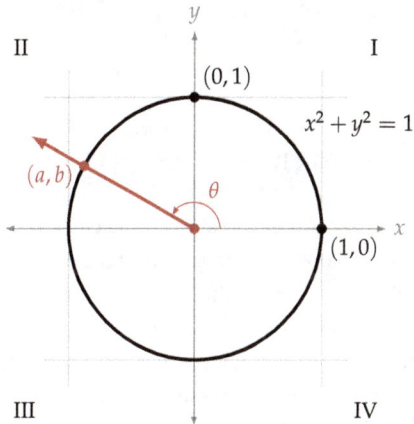

Therefore we know that for any angle θ:

$$-1 \leq \cos\theta \leq 1,$$

$$-1 \leq \sin\theta \leq 1.$$

LESSON 5

THE COMPLEMENTARY ANGLE THEOREM

While less obvious with our new definitions, the following identities continue to hold for any angle θ, provided that the trigonometric values are defined.

$$\sin\theta = \cos(90° - \theta) \qquad \cos\theta = \sin(90° - \theta)$$

$$\csc\theta = \sec(90° - \theta) \qquad \sec\theta = \csc(90° - \theta)$$

$$\tan\theta = \cot(90° - \theta) \qquad \cot\theta = \tan(90° - \theta)$$

THE EVEN/ODD PROPERTIES

The next property could not have been discussed before this lesson, as we had not yet defined what it means for an angle to be negative. To illustrate the property, consider an angle θ along with its negative. Both angles are shown in the Unit Circle to the right. Recall that positive angles are measured counterclockwise from the positive x-axis, while negative angles are measured clockwise.

We see that the points corresponding to these two angles have the same x-coordinate $x = a$. This will be true regardless of the angle θ chosen, or the quadrant in which its terminal side lies.

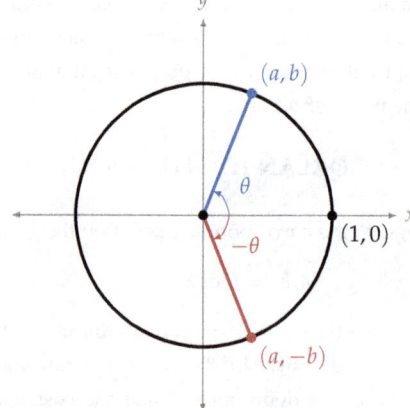

Since these x-coordinates are the cosines of their respective angles, we know that

$$\cos(-\theta) = \cos(\theta).$$

On the other hand, the y-coordinates of the respective points are each other's negatives, for any choice of θ:

$$\sin(-\theta) = -\sin(\theta).$$

These relationships are often expressed by saying that cosine is an **even** function, and sine is an **odd** function.

EXERCISES

For Exercises 1–9, state whether the value of each expression is positive or negative.

1. $\sin 300°$
2. $\cos(-210°)$
3. $\sin(-270°)$
4. $\sin\left(-\dfrac{4\pi}{3}\right)$
5. $\cos\left(-\dfrac{3\pi}{4}\right)$
6. $\sin\left(\dfrac{7\pi}{3}\right)$
7. $\tan\left(\dfrac{11\pi}{6}\right)$
8. $\cot(-751°)$
9. $\csc\left(\dfrac{6\pi}{7}\right)$

For Exercises 10–18, find the exact value of each expression, without using a calculator.

10. $\sin 240°$
11. $\cos 150°$
12. $\cos(-150°)$
13. $\cos\left(\dfrac{11\pi}{3}\right)$
14. $\sin 900°$
15. $\sin\left(-\dfrac{\pi}{6}\right)$
16. $\sec 1560°$
17. $\tan\left(-\dfrac{5\pi}{3}\right)$
18. $\cot 390°$

19. Suppose that the terminal side of θ lies in Quadrant II, and that $\sin\theta = \dfrac{8}{17}$. What is $\tan\theta$?

20. Suppose that the terminal side of θ lies in Quadrant IV, and that $\cos\theta = \dfrac{5}{13}$. What is $\cot\theta$?

21. Find the exact value of:
$$\sin 1° + \sin 2° + \sin 3° + \cdots + \sin 358° + \sin 359°.$$

22. Find the exact value of:
$$\cos 1° + \cos 2° + \cos 3° + \cdots + \cos 358° + \cos 359°.$$

23. If $\cos\theta = \dfrac{4}{5}$, find the exact value of $\sec(\theta + \pi)$.

24. If $\sin\theta = \dfrac{6}{7}$, find the exact value of $\csc(\theta + \pi)$.

Use a sketch of the Unit Circle to determine which of $\sin\theta$ or $\tan\theta$ has the larger value, when:

25. θ is in Quadrant I
26. θ is in Quadrant II
27. θ is in Quadrant III
28. θ is in Quadrant IV

29. Recall that sine is an **odd** function, and cosine is an **even** function, meaning that
$$\sin(-\theta) = -\sin(\theta),$$
$$\cos(-\theta) = \cos(\theta).$$
What about the remaining four trigonometric functions? Determine if each function is even, odd, or neither.

LESSON 6

Functions and Inverse Functions

Early in Lesson 5 we made reference to the fact that we'd start thinking about sine and cosine as **functions**, without being very specific about exactly what a function is, or why it should be important. It is time to remedy this.

Painting with a very broad brush, a function is just a rule that takes a given object (an **input**) and assigns it a unique value (the **output**). We might think about the price that is given to cup of coffee in a coffee shop, or the social security number that is given to each resident of the United States. A cup of coffee should not have two different prices at once, nor should a single person have two different social security numbers. This property is what we look for in functions.

We could also think of **measurement** itself as a function, assuming that we were constrained to use a fixed unit. For example, each side of this page has a length in centimeters, and the act of measuring each of these lengths assigns each of them a numerical value that is unique to that side. A side cannot have more than one length.

Another function is described by the act of **converting** from one unit to another. In the above example, for each length that we have in centimeters, there is another, unique numerical value that describes that same length in inches. We even know how to find this new value, using the dimensional analysis that we learned in Lesson 2.

Lesson 6

Example: Suppose that the length of a steel bar is 250 cm. Because we know that 1 cm = 0.393701 in, we can find the equivalent length by multiplication:

$$\text{length} = 250\,\text{cm} \cdot \frac{0.393701\,\text{in}}{1\,\text{cm}} = 98.42525\,\text{in}.$$

We'll write this in an alternative notation, without the units:

$$\text{len}(250) = 0.393701\,(250)$$
$$= 98.42525.$$

The expression on the left is **not a multiplication**. There is no object here named "len" that can be multiplied by the real number 250. Here, "len" is the name of the rule that assigns to each (centimeter) length its corresponding (inch) length. The description of how this function works is written on the right.

If the initial measurement in centimeters was not known, we can still describe this function using a blank variable:

$$\text{len}(x) = 0.393701\,(x).$$

Once again, the expression on the left is the name of the function, not a multiplication. The expression on the right **is** a multiplication, and defines how the function works.

All of the unit conversions that we have seen so far have been defined by multiplication, using dimensional analysis. This method works for nearly all units. But occasionally, the relationships between units are a bit more complex. One prototypical example is the relationship between degrees of temperature, which are commonly measured in either degrees Fahrenheit or degrees Celsius.

Definition: Given a temperature measurement of x in degrees Celsius, we convert it to Fahrenheit via the formula

$$\text{far}(x) = \frac{9}{5}x + 32.$$

On the left we see the name of this function, along with the name of the variable that represents the function's input. On the right we see the function's definition, which involves a multiplication followed by an addition.

For example, converting 25°C would return the value

$$\text{far}(25°\text{C}) = \frac{9}{5}\,(25) + 32 = 9(5) + 32 = 77°\text{F}.$$

Dimensional analysis will not work for us in this case, as our formula is more complex than the simple multiplication that we saw in the function $\text{len}(x)$. Nevertheless, the rule returns a unique output for every temperature given to it.

LESSON 6

As we might imagine, there is a corresponding formula to convert from Celsius to Fahrenheit:

Definition: Given a temperature measurement of x in degrees Fahrenheit, we convert it to Celsius via the formula

$$\text{cel}(x) = \frac{5}{9}(x - 32).$$

In other words, this new function acts by first adding -32, and then multiplying by $5/9$, which are the additive and multiplicative inverses of the constants that appeared in $\text{far}(x)$, respectively. Coupled with the fact that these operations occur in the opposite order, we should expect that each of these functions should "undo" the other.

We can test this by evaluating our new function at $77°F$:

$$\text{cel}(77°F) = \frac{5}{9}(77 - 32)$$
$$= \frac{5}{9}(45)$$
$$= 25°C.$$

This was the original degree measure that we started with, as we expected. Of course, we've tested this only at a single point, but we will address this problem in more detail soon.

But before we move forward, we would be well served to acknowledge one of the most common mistakes associated with functions. We've already mentioned several times that the notation we use for functions does **not** indicate a multiplication. Therefore we should **not** expect functions to satisfy the distributive property.

Example: Consider the function $\text{far}(x)$, and suppose that we have two temperatures, $x = 14°C$ and $y = 21°C$. Then:

$$\text{far}(x + y) = \text{far}(14°C + 21°C)$$
$$= \frac{9}{5}(35) + 32$$
$$= 9(7) + 32$$
$$= 95°F.$$

On the other hand,

$$\text{far}(x) + \text{far}(y) = \text{far}(14°C) + \text{far}(21°C)$$
$$= \left[\frac{9}{5}(14) + 32\right] + \left[\frac{9}{5}(21) + 32\right]$$
$$= \frac{9}{5}(35) + 64$$
$$= 9(7) + 64$$
$$= 127°F.$$

LESSON 6

We've just found that

$$\text{far}(x+y) \neq \text{far}(x) + \text{far}(y).$$

This may seem to be at odds with true statements like one that we've just seen,

$$\frac{9}{5}(35) = \frac{9}{5}(14+21) = \frac{9}{5}(14) + \frac{9}{5}(21),$$

but we must continue to remind ourselves that "far" is not a real number, and therefore it does not behave like one.

So far in this lesson we've been ignoring the trigonometric functions. Each of these represents a sort of rule that links angles with real numbers. For example, given an angle θ, the sine function performs the following actions:

- Draws the angle θ in the x, y-plane, with its initial side on the x-axis.
- Finds the point P where the terminal side of θ intersects the Unit Circle.
- Returns the y-coordinate of P.

This definition of the value $\sin(\theta)$ is certainly more complex than a simple multiplication.

Example: Consider the function $\sin(\theta)$, and suppose that we have two angles, $\alpha = 60°$ and $\beta = 120°$. Then:

$$\sin(\alpha + \beta) = \sin\left(\frac{\pi}{3} + \frac{2\pi}{3}\right) = \sin(\pi) = 0.$$

However, evaluating these individually gives

$$\sin(\alpha) + \sin(\beta) = \sin\left(\frac{\pi}{3}\right) + \sin\left(\frac{2\pi}{3}\right)$$
$$= \frac{\sqrt{3}}{2} + \frac{\sqrt{3}}{2}$$
$$= \sqrt{3}.$$

This "non-additivity" holds for the functions $\sin(\theta)$ and $\text{far}(x)$, but is also true for **almost all functions**, provided that they have even a small amount of complexity to them.

Now that we've seen a few examples and issued our most important warning, it's time to learn a bit more about the language of functions. This has a very particular vocabulary that we'll be using in the upcoming lessons.

Definition: Let f be any function. The set of all possible inputs for which f provides an output is called the **domain** of the function. The set of all values that are output by f is called the **range** of the function.

Example: Consider the cosine function, $\cos(\theta)$.

We know that any angle can be input to cosine, including negative angles and angles larger than 2π. Since these angles are each measured by a real number, we conclude that the **domain** of $\cos(\theta)$ is all of $\mathbb{R}$.

However, because $\cos(\theta)$ represents the x-coordinate of a point on the Unit Circle, the values of $\cos(\theta)$ can be at maximum equal to 1, and at minimum -1. That is, all values of $\cos(\theta)$ fall within the **range** $[-1,1]$.

Example: Consider the tangent function, $\tan(\theta)$.

Because tangent is defined as the ratio of sine to cosine,

$$\tan(\theta) = \frac{\sin(\theta)}{\cos(\theta)},$$

there are restrictions on the angles θ that may be input to this function. Specifically, θ cannot be equal to any angle that makes $\cos(\theta) = 0$. There are many of these, and we exclude them when we write the **domain** of tangent below:

$$\mathbb{R} \setminus \left\{ \ldots, -\frac{3\pi}{2}, -\frac{\pi}{2}, \frac{\pi}{2}, \frac{3\pi}{2}, \ldots \right\}$$

For the range of tangent, first consider the interval $[0°, 90°)$ that corresponds to the first quadrant.

The values of tangent are positive on this interval, and we also know that $\tan(0) = 0$. On the other end of the interval, as θ gets closer to $90°$, we can use a calculator to estimate the values of tangent:

$$\tan(89°) \approx 57.29$$
$$\tan(89.5°) \approx 114.59$$
$$\tan(89.75°) \approx 229.18$$
$$\tan(89.99°) \approx 5729.58.$$

Unlike sine and cosine, the values of tangent are not bounded above by 1, nor by any other value. We can continue to enter angles still closer to $90°$ to make the value of tangent as large as we like.

Furthermore, tangent is an **odd function**, meaning that inputting the negative of any of these angles will return the corresponding negative value:

$$\tan(-89.99°) \approx -5729.58.$$

We conclude that the values of tangent can fall anywhere in the **range** $\mathbb{R}$. In fact, we can achieve any of these values just by using angles in the interval $(-90°, 90°)$. This is an important point that will be used again in Lesson 7.

Lesson 6

Functions are often defined directly by a formula, leaving it to the reader to determine their domain and range. However, the domain and range can be specified in the definition of a function by using the notation below. We use cosine for this example, with the relevant parts of the notation labeled.

name of the function   domain   range

$$\cos : \mathbb{R} \longrightarrow [-1, 1]$$
$$\theta \longmapsto \cos\theta$$

variable   definition of the function

As another example, we just saw that the tangent function takes all of the values in its range $\mathbb{R}$ using only the angles $\theta \in (-90°, 90°)$. This defines a complete function:

$$\tan : (-90°, 90°) \longrightarrow \mathbb{R}$$
$$\theta \longmapsto \tan\theta.$$

This notation is a bit more cumbersome than simply saying $\tan\theta$, but it removes the guesswork from trying to establish a domain and range. It also emphasizes one important element of a complete function:

Definition: The specific variable or expression that is input to a given function is called the **argument** of the function.

For trigonometric functions, this term refers to the angle at which the function is evaluated, usually called θ.

For more general functions, the term "argument" is usually used when the function's input is something more complex than a single variable.

Example: Suppose that f is the function defined by

$$f : \mathbb{R} \longrightarrow [2, \infty)$$
$$x \longmapsto x^2 + 2.$$

Find and simplify the expression $f(z^3 - 1)$.

When asked to find $f(z^3 - 1)$ the expression $z^3 - 1$ is the **argument** of f. This is what we will be inputting to the definition of f before we simplify:

$$f(x) = x^2 + 2$$
$$f(z^3 - 1) = (z^3 - 1)^2 + 2$$
$$= (z^6 - 2z^3 + 1) + 2$$
$$= z^6 - 2z^3 + 3.$$

The idea of a function having a somewhat complex argument leads to another important definition. Note that the argument $z^3 - 1$ could be considered a function in its own right, which we might write as

$$g : \mathbb{R} \longrightarrow \mathbb{R}$$
$$x \longmapsto z^3 - 1.$$

In this context, what we did in the last example amounts to inputting one function to another. This process is known as **composition**.

Definition: Suppose that f and g are two functions. Then the **composition of f with g** is

$$(f \circ g)(x) = f(g(x)).$$

Example: Consider the functions

$$f(x) = -3x^2 + 7x - 2,$$
$$g(x) = x^2 - 3.$$

We'll use these functions to create the composition $f \circ g$ as described above. But note that a different composition can also be formed by plugging the function f into g:

$$(g \circ f)(x) = g(f(x)).$$

We will actually find both of these compositions.

First comes the composition $f \circ g$, found by inputting g to f and then simplifying:

$$(f \circ g)(x) = f(g(x))$$
$$= f(x^2 - 3)$$
$$= -3(x^2 - 3)^2 + 7(x^2 - 3) - 2$$
$$= -3(x^4 - 6x^2 + 9) + 7(x^2 - 3) - 2$$
$$= -3x^4 + 18x^2 - 27 + 7x^2 - 21 - 2$$
$$(f \circ g)(x) = -3x^4 + 25x^2 - 50.$$

Then we find $g \circ f$ in a similar way:

$$(g \circ f)(x) = g(f(x))$$
$$= g(-3x^2 + 7x - 2)$$
$$= (-3x^2 + 7x - 2)^2 - 3$$
$$= \left(9x^4 - 42x^3 + 61x^2 - 28x + 4\right) - 3$$
$$(g \circ f)(x) = 9x^4 - 42x^3 + 61x^2 - 28x + 1.$$

If we think of functions as mathematical objects in their own right, then composition defines an **operation** on these functions; a way to take two given functions and combine them to obtain a new one.

LESSON 6

This operation on functions is very similar to the addition and multiplication that are defined on real numbers. Each of these also gives us a way to combine two objects into a new one. However, as we've just seen, composition of functions is not **commutative** in the way that addition and multiplication are.

Nevertheless, there is more to this analogy. Remember that both addition and multiplication have **identities**. Being 0 and 1 respectively, these satisfy the following properties for every nonzero real number a:

$$a + 0 = a,$$
$$a \cdot 1 = a.$$

There is also a **function** that satisfies a similar property with respect to function composition.

Definition: The **identity function**, denoted $\text{id}(x)$, is the function that sends every real number to **itself**:

$$\text{id} : \mathbb{R} \longrightarrow \mathbb{R}$$
$$x \longmapsto x.$$

To see the relevance of this function, let's return to the functions $\text{far}(x)$ and $\text{cel}(x)$ that we saw earlier.

Those two functions are connected by the fact that they "undo" each other. That is, we saw that

$$\text{far}(25°\text{C}) = 77°\text{F},$$

and later we found that

$$\text{cel}(77°\text{F}) = 25°\text{C}.$$

With our new definitions we can rewrite these statements as a pair of compositions:

$$\text{far}\big(\text{cel}(77°\text{F})\big) = 77°\text{F},$$
$$\text{cel}\big(\text{far}(25°\text{F})\big) = 25°\text{F}.$$

Furthermore, we suspect that these compositions will work this way for **any** temperature that we start with. That is,

$$(\text{far} \circ \text{cel})(x) = x,$$
$$(\text{cel} \circ \text{far})(x) = x.$$

Of course, if we pay attention to units, the domain and range of the first composition consist of temperatures expressed in degrees Fahrenheit, while the domain and range of the second consist of temperatures in degrees Celsius. Still, each represents the **identity function** on its domain.

Definition: Let $f : D \longrightarrow R$ be a function with domain D and range R. Suppose also that there exists a function $g : R \longrightarrow D$ with the properties that

$$g \circ f : D \longrightarrow D \qquad f \circ g : R \longrightarrow R$$
$$x \longmapsto x \qquad\qquad x \longmapsto x$$

That is, $g \circ f = $ id on D, and $f \circ g = $ id on R. Then we call g the **inverse** of f, and write $g = f^{-1}$.

It should be clear from these definitions that if g is the inverse of a function f, then f is also the inverse of g, and we could also write $f = g^{-1}$.

Example: We can confirm our suspicion that far(x) and cel(x) are inverse to each other by verifying this definition. We first check the composition far $\circ$ cel:

$$\text{far}(\text{cel}(x)) = \text{far}\left(\frac{5}{9}(x-32)\right)$$
$$= \frac{9}{5}\left(\frac{5}{9}(x-32)\right) + 32$$
$$= (x-32) + 32$$
$$\text{far}(\text{cel}(x)) = x.$$

We then check the composition cel $\circ$ far:

$$\text{cel}(\text{far}(x)) = \text{cel}\left(\frac{9}{5}x + 32\right)$$
$$= \frac{5}{9}\left[\left(\frac{9}{5}x + 32\right) - 32\right]$$
$$= \frac{5}{9}\left[\frac{9}{5}x\right]$$
$$\text{cel}(\text{far}(x)) = x.$$

Therefore each composition represents the identity function on its respective domain. This verifies that far(x) is the inverse of cel(x), and vice versa.

The definition we're using also seems to imply that not all functions have inverses. This is indeed the case. We don't even have to look very far for an example.

Example: Let f be the sine function: $f(\theta) = \sin\theta$.

This function has the same domain and range as the cosine function. We established this earlier to be

$$f : \mathbb{R} \longrightarrow [-1, 1].$$

However, like the cosine function, sine **repeats its values**.

LESSON 6

Suppose that we choose a specific value from the range of sine, like $y = 0$. The Unit Circle shows us that there are many angles that give this same value, such as $\theta = 0$, or $\theta = \pi$, or $\theta = 2\pi$, just to name a few.

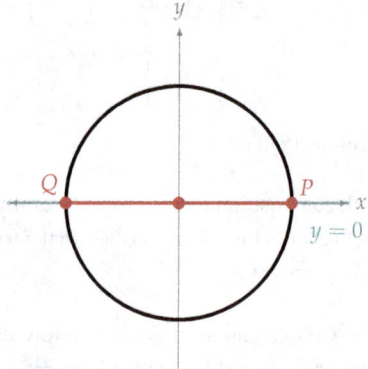

If the sine function were to have an inverse, then the domain of this inverse should be the range of f, and the range of the inverse should be the domain of f, so that

$$f^{-1}: [-1, 1] \longrightarrow \mathbb{R}.$$

With regard to units, the domain of f^{-1} would be an interval of real, unitless values, and the range would be a set of angles expressed in either radians or degrees.

With a collection of arguments that return a value of zero,

$$f(0) = 0$$
$$f(\pi) = 0$$
$$f(2\pi) = 0$$

we can rewrite each of them in terms of the inverse function:

$$0 = f^{-1}(0)$$
$$\pi = f^{-1}(0)$$
$$2\pi = f^{-1}(0).$$

But this implies that $f^{-1}(0)$ is equal to at least three different things. This violates the very property that would define f^{-1} as a **function**; that every input should return a unique output.

This is the criterion that we'll look for when determining whether or not a function has an inverse; that no two distinct inputs return the same value.

Definition: A function $f(x)$ is called **one-to-one** if, for every two distinct points $x \neq y$, we know that

$$f(x) \neq f(y).$$

LESSON 6

We've seen already that **all** of the trigonometric functions repeat themselves periodically, as coterminal angles always have the same trigonometric values. This implies that **none** of the trigonometric functions are one-to-one. So how are we to define their inverse functions?

The key to this is **restricting** the domains of the trigonometric functions to regions on which they **are** one-to-one.

For the sine function $f(\theta) = \sin\theta$ that we were just working with, we'd like to find an interval of angles that is large enough that sine takes all of the values in the range $[-1, 1]$, but small enough so that they don't repeat.

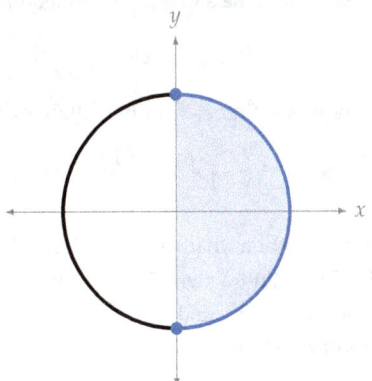

Because sine represents the y-coordinate of points on the Unit Circle, we can single out just one half of the circle that provides all of these y-coordinates. We'll try to keep the angles as small as possible when doing this. We then define the "restricted" sine function, which is one-to-one on this restricted domain.

$$\sin : \left[-\frac{\pi}{2}, \frac{\pi}{2}\right] \longrightarrow [-1, 1]$$
$$\theta \longmapsto \sin\theta$$

This effectively means that of all of the possible values for $f^{-1}(0)$ that we saw earlier, we are selecting one of them as the **principal** value.

Definition: The restricted domain on which a function f becomes one-to-one is called the **fundamental domain** of f. The choice of fundamental domain is rarely unique. However, f will always have an inverse function on such a fundamental domain.

For every x in the range of f, the unique value of $f^{-1}(x)$ that lies in the fundamental domain is called the **principal value** of $f^{-1}(x)$.

We will spend the next lesson determining the fundamental domains for each of the six trigonometric functions.

LESSON 6

EXERCISES

For each function f, find the values $f(2)$, $f(-2)$, and $f(0)$.

1. $f(x) = x^2 + 4$
2. $f(x) = |x - 2|$
3. $f(x) = \dfrac{x^2}{3} - \dfrac{x^3}{2}$
4. $f(x) = x^4 - 6$
5. $f(x) = |x| - 2$
6. $f(x) = \sin\left(\dfrac{\pi}{4}x\right)$

For each pair of functions, find $(f \circ g)(3)$ and $(g \circ f)(3)$.

7. $f(x) = x + 2$, $g(x) = x - 1$
8. $f(x) = x^3 - 1$, $g(x) = x + 1$
9. $f(x) = x^2 + 8$, $g(x) = x - 3$
10. $f(x) = x^2$, $g(x) = x^3$

For each pair of functions, find $(f \circ g)(x)$ and $(g \circ f)(x)$.

11. $f(x) = 2x + 1$, $g(x) = x - 3$
12. $f(x) = x^2 + 3$, $g(x) = 2x - 1$
13. $f(x) = -x^2 - 8$, $g(x) = x^2 - 1$
14. $f(x) = x + 2$, $g(x) = x - 2$

For each pair of functions, determine whether or not f and g are inverse to each other on their respective domains.

15. $f(x) = 2x + 1$, $g(x) = \dfrac{x - 1}{2}$
16. $f(x) = -2x + 3$, $g(x) = 2x - 3$
17. $f(x) = \dfrac{5}{x - 2}$, $g(x) = \dfrac{x - 2}{5}$
18. $f(x) = 4 + x^2$, $g(x) = \pm\sqrt{x - 4}$

In Exercises 19–21, let f be a one-to-one function having
$$f(0) = -1, \quad f(\pi) = 1, \quad \text{and } f\left(\dfrac{\pi}{3}\right) = \dfrac{1}{2}.$$
With these conditions, find each of the following values:

19. $f^{-1}(-1)$
20. $f^{-1}\left(\dfrac{1}{2}\right) + f^{-1}\left(\dfrac{1}{2}\right)$
21. $f^{-1}\left(\dfrac{1}{2} + \dfrac{1}{2}\right)$

In Exercises 22–24, show that each function f is **not** one-to-one on $\mathbb{R}$. Then find a new domain for each function on which it **is** one-to-one, and write the new f using the notation introduced in this lesson.

22. $f(x) = x^4$
23. $f(x) = x^2 - 4$
24. $f(x) = (x - 4)^2$

LESSON 7

The Inverse Trigonometric Functions

Here we'll be using what we learned in Lesson 6 to define inverses for the six trigonometric functions. These functions are not one-to-one, so it will be necessary to restrict their domains to smaller intervals. In this context, we call these the **fundamental domains** of the trigonometric functions.

The result of this process will be a set of six new functions, with carefully selected domains and ranges. These are appropriately called Inverse Trigonometric Functions, though we sometimes refer to them as the **arc-functions**. Their domains (inputs) will be sets of pure, dimensionless, unitless real numbers. Their ranges (outputs) will be angles, usually expressed in radians, though not necessarily.

Each fundamental domain will create a set of **principal values** for its respective inverse function. Given the context, these values are sometimes called **principal angles**.

Throughout this lesson we'll be evaluating these inverse functions, and we will expect the principal angles we find to be unique. In the next lesson we will approach the more general question of finding **all** possible angles satisfying an **equation**. This will usually involve finding the principal angle first, and then generating the solutions from it.

We first approach the function $f(\theta) = \sin\theta$, which represents the y-coordinates of the points on the Unit Circle.

LESSON 7

Definition: We choose the fundamental domain of the sine function to be the interval

$$D = \left[-\frac{\pi}{2}, \frac{\pi}{2}\right].$$

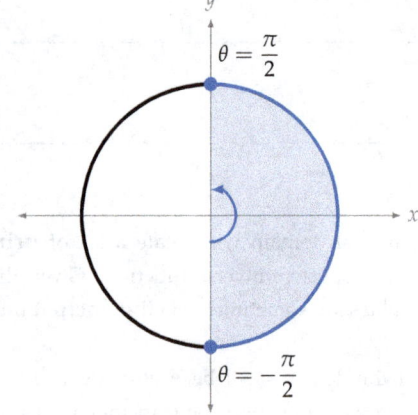

On this domain, the function $f(x) = \sin x$ has range $[-1, 1]$, so we can define **the inverse sine function**:

$$f^{-1} \colon [-1, 1] \longrightarrow \left[-\frac{\pi}{2}, \frac{\pi}{2}\right]$$

$$x \longmapsto \sin^{-1} x$$

Example: Let $g(x) = \sin^{-1} x$. Find $g\left(\frac{1}{2}\right)$.

Knowing that the values output by g should be angles, we first attempt to find **all** of the angles at which $\sin \theta = 0.5$. The sine of an angle corresponds to its y-coordinate, so we look on the Unit Circle for points having $y = 0.5$. There are two such points, labeled P and Q below.

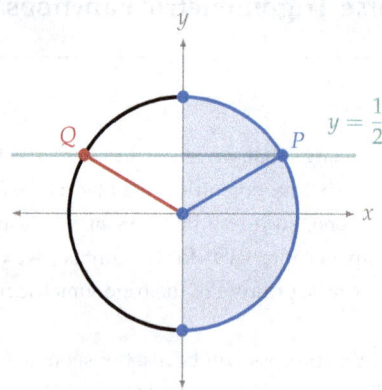

We recognize this height as one coming from our standard triangles; namely, it is the short side of a 30/60/90 triangle. Therefore both of these points have reference angles of 30° with the x-axis.

There are an infinite number of coterminal angles that could correspond to these points. We list just a few below:

$$\theta \in \left\{\ldots, -\frac{11\pi}{6}, -\frac{7\pi}{6}, \frac{\pi}{6}, \frac{5\pi}{6}, \frac{13\pi}{6}, \frac{17\pi}{6}, \ldots\right\}.$$

However, among all of these is exactly one angle that lies in the fundamental domain of sine, which is shaded in blue. That angle is $\theta = 30°$, which is the unique output of this inverse function:

$$g\left(\frac{1}{2}\right) = \sin^{-1}\left(\frac{1}{2}\right) = \frac{\pi}{6}.$$

This is the **principal value** of $g(x) = \sin^{-1} x$ at this point, and is the value that will be returned by a calculator if we had used one.

The fundamental domain of sine consists of two quadrants: Quadrants I and IV. The angles in Quadrant I are positive, and likewise the values of sine in this quadrant are also positive. Conversely, the angles in Quadrant IV are negative, along with the values of sine. This has everything to do with the fact that sine is an **odd** function.

So looking at the **sign** of the argument x tells us something about the principal value of $g(x)$, and the quadrant in which we should expect it to lie. And in between these two quadrants, our definitions preserve one of the nicest properties that the sine function has: the fact that $\sin(0) = 0$.

Next we'll say a word about the **notation** we use for these functions. The "negative exponent" notation we've just used is inherited from what is used for general inverse functions. However, this can sometimes be problematic, as it is also used for **reciprocals**, or multiplicative inverses. To avoid confusion we often use the following notations:

$$\arcsin x = \sin^{-1}(x) \qquad \arccos x = \cos^{-1}(x)$$
$$\text{arccsc}\, x = \csc^{-1}(x) \qquad \text{arcsec}\, x = \sec^{-1}(x)$$
$$\arctan x = \tan^{-1}(x) \qquad \text{arccot}\, x = \cot^{-1}(x)$$

The arc-functions are so named because their outputs are often given as radian measures, which double as the length of an arc on the Unit Circle.

Of course, we haven't yet defined most of these functions, or their domains and ranges. But we will work on this now.

Next up is the cosine function. Cosine is also not one-to-one on its domain, so we'll again need to choose a fundamental domain for it. This time it will be possible to keep all of the angles **positive**, which we could not do with sine.

LESSON 7

Definition: We choose the fundamental domain of the cosine function to be the interval

$$D = [0, \pi].$$

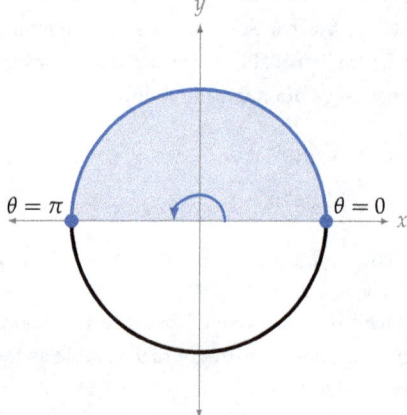

On this domain the function $f(x) = \cos x$ has range $[-1, 1]$, so we can define **the inverse cosine function**:

$$f^{-1}: [-1, 1] \longrightarrow [0, \pi]$$

$$x \longmapsto \arccos x$$

Example: Find the exact value of

$$\sin\left(\arccos\left(-\frac{\sqrt{3}}{2}\right)\right).$$

This example is actually a composition of two different functions: arccosine and sine. The arc-function is being evaluated first, since it is written on the right, and its output will be an **angle** in the interval $[0, \pi]$.

Specifically, the cosine of this angle corresponds to the x-coordinate of a point on the Unit Circle, and there are two points at which $x = -\sqrt{3}/2$.

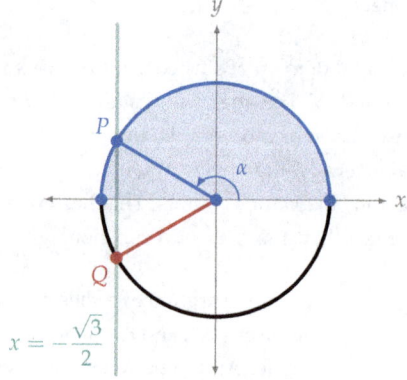

Of course, only P creates an angle α that lies in the fundamental domain of cosine, so this angle is the unique output of the expression
$$\alpha = \arccos\left(-\frac{\sqrt{3}}{2}\right).$$

To find the sine of this angle we'll need to find the y-coordinate of P, but we already know quite a bit about this point. First of all, its y-coordinate is **positive**. Also, it is a vertex of the right triangle with a horizontal side of length $\sqrt{3}/2$ and hypotenuse 1, so we can use the Pythagorean Theorem to find the length of the vertical side:
$$y = \sqrt{(1)^2 - \left(\frac{\sqrt{3}}{2}\right)^2} = \sqrt{1 - \frac{3}{4}} = \frac{1}{2}.$$

We now conclude that
$$\sin\left(\arccos\left(-\frac{\sqrt{3}}{2}\right)\right) = \frac{1}{2}.$$

Note also that it wasn't necessary to actually find α to solve the problem; we only needed a point on the Unit Circle that takes into account both the input value and the fundamental domain of the functions given. Though, in this instance, it would not be difficult to find the angle as well.

The two fundamental domains that we've just seen were both **closed**, meaning that they contained their endpoints. Unfortunately, the fundamental domains for the four remaining trigonometric functions will not be as convenient.

We'll start with the fundamental domain of tangent. Recall that tangent is defined as the ratio of sine to cosine:
$$\tan\theta = \frac{\sin\theta}{\cos\theta}.$$

In particular, this means that the domain of tangent does not include any angles that would make the denominator equal to zero, meaning that
$$\theta \notin \left\{\ldots, -\frac{3\pi}{2}, -\frac{\pi}{2}, \frac{\pi}{2}, \frac{3\pi}{2}, \ldots\right\}.$$

Luckily, the tangent function is both defined and one-to-one on the intervals between each of these angles. One of these intervals is the same fundamental domain as the sine function, excluding the endpoints: $(-90°, 90°)$.

Choosing this interval lets us use a good deal of our previous work. Cosine is always positive on this interval, but the values of sine change sign depending on the sign of the angle. Therefore tangent is also **odd** on its fundamental domain, and also has the property that $\tan(0) = 0$.

LESSON 7

Definition: We choose the fundamental domain of the tangent function to be the interval

$$D = \left(-\frac{\pi}{2}, \frac{\pi}{2}\right).$$

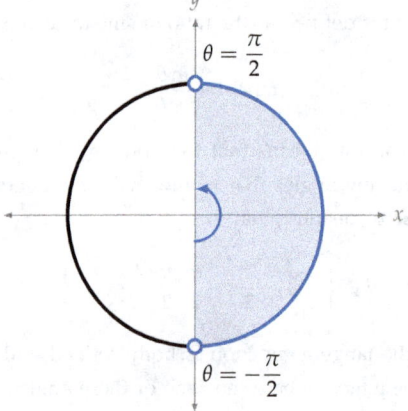

On this domain the function $f(x) = \tan x$ has range $\mathbb{R}$, so we can define **the inverse tangent function**:

$$f^{-1}: (-\infty, \infty) \longrightarrow \left(-\frac{\pi}{2}, \frac{\pi}{2}\right)$$

$$x \longmapsto \arctan x$$

Example: Find the exact value of both of the following:

$$\arctan\left(\cos\left(-\frac{\pi}{3}\right)\right),$$

$$\arccos\left(\tan\left(-\frac{\pi}{3}\right)\right).$$

These expressions are both compositions, as we saw in the last example, though this time the functions are being composed in the opposite order. We start by evaluating the first expression exactly as it is given.

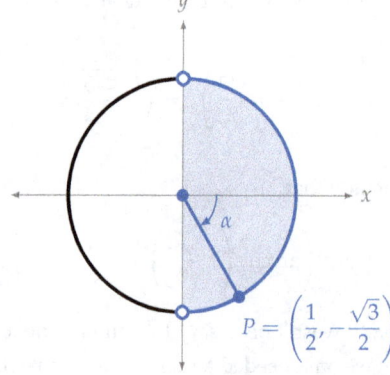

We're starting from a specific angle, $\alpha = -60°$. This creates a 30/60/90 triangle in Quadrant IV, shown above.

The vertex of this triangle is labeled P, and its x-coordinate is the value of cosine for α:

$$\cos\left(-\frac{\pi}{3}\right) = \frac{1}{2}.$$

Now we'll need to look for angles β at which

$$\tan \beta = \frac{1}{2}.$$

This is a positive value, so we should expect β to be positive as well, thus in Quadrant I. But since tangent is a **ratio** of x and y-values, this point is not as easy to see on the Unit Circle as it was for sine and cosine. Also, this value does not appear for the tangent of any of the standard angles. Without many more options, we turn to a calculator, which tells us that:

$$\arctan(0.5) \approx 26.57°.$$

Turning to the second expression, we first evaluate tangent at the given angle:

$$\tan\left(-\frac{\pi}{3}\right) = \frac{-\sqrt{3}/2}{1/2} = -\sqrt{3}.$$

Just as before, we'll now try to find angles β at which

$$\cos \beta = -\sqrt{3}.$$

But this is a problem, because $-\sqrt{3} \approx -1.732$, which is outside the range of cosine's values.

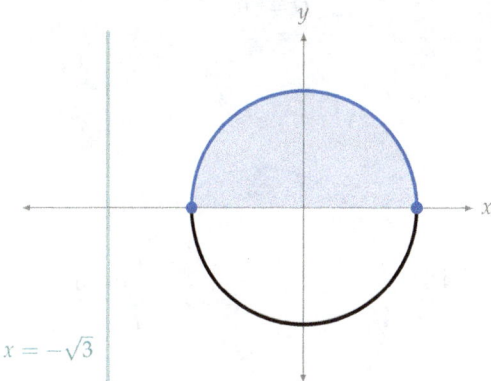

Therefore no such angle β exists. In this situation, when

$$\beta = \arccos\left(\tan\left(-\frac{\pi}{3}\right)\right),$$

then we say that both the angle β and the expression on the right are **undefined**.

The next function in line is cotangent, which has a definition similar to tangent, and therefore will have a similar fundamental domain.

LESSON 7

Definition: We choose the fundamental domain of the cotangent function to be the interval

$$D = (0, \pi).$$

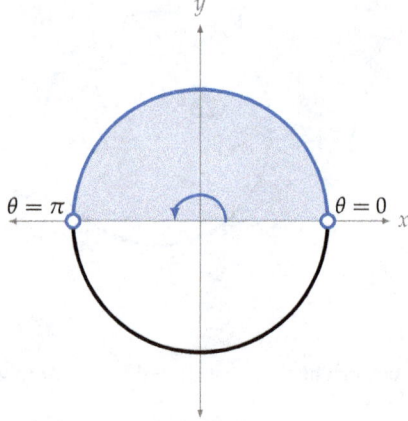

On this domain the function $f(x) = \cot x$ has range $\mathbb{R}$, so we define **the inverse cotangent function**:

$$f^{-1}: (-\infty, \infty) \longrightarrow (0, \pi)$$

$$x \longmapsto \operatorname{arccot} x$$

Again, since the cotangent function is defined as the ratio of cosine to sine

$$\cot \theta = \frac{\cos \theta}{\sin \theta},$$

we'll need avoid any angles that would make the denominator equal to zero. Here these are

$$\theta \notin \{\ldots, -\pi, 0, \pi, 2\pi, 3\pi, \ldots\}.$$

But on the intervals between these angles, the cotangent function is one-to-one just as the tangent function was. Therefore we've been able to match its fundamental domain with one that we've already seen: cosine.

Example: Find the exact value of

$$\sin(\operatorname{arccot}(-1)).$$

This composition is written with the arc-function being applied first. We'll need to establish the angles α for which

$$\cot \alpha = -1,$$

or equivalently, for which $\sin \alpha = -\cos \alpha$. On the Unit Circle this occurs for two angles, $135°$ and $315°$. The former lies within the fundamental domain of cotangent, and so is the principal value of the arccotangent function.

LESSON 7

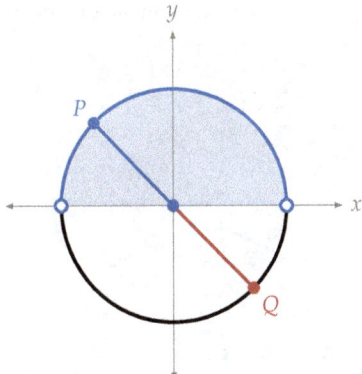

This angle is in Quadrant II, where cotangent is negative. However, the values of sine are positive in this quadrant, so we conclude that

$$\sin(\text{arccot}(-1)) = \sin(135°) = \frac{\sqrt{2}}{2}.$$

Example: Find an algebraic expression, not involving any trigonometric expressions, that is equivalent to

$$\cos(\arcsin x).$$

Here we are asked to manipulate an expression without knowing anything about the argument x, aside from the fact that it must be in the interval $[-1, 1]$, the domain of arcsine. As a first step, we simply give a name to the angle that is the arcsine of x:

$$\theta = \arcsin x.$$

Just this simple act gives us a relationship between θ and x that we can rewrite as

$$x = \sin\theta,$$

and also express in a picture that shows the relationship between the angle θ and the length of the opposite side of a right triangle x:

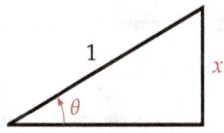

From what we know about the triangle we can fill in the third side with the Pythagorean Theorem:

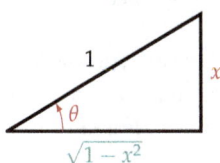

79

Lesson 7

With this triangle in place we can evaluate any of the six trigonometric functions that we like at the angle θ, despite the fact that we don't know what θ is. In the example we're asked to find

$$\cos\theta = \cos(\arcsin x),$$

and as a ratio of the adjacent side of this triangle to the hypotenuse, we find it to be

$$\cos(\arcsin x) = \frac{\sqrt{1-x^2}}{1} = \sqrt{1-x^2}.$$

Note that the expression we've just found does not involve any trigonometric functions or their inverses. Also, both this expression and the arcsine function itself have the **same domain**: $[-1, 1]$.

The two arc-functions that remain are far less commonly used than those that we've already seen. In fact, most hand calculators will not even have buttons that allow us to calculate the arcsecant or arccosecant of a given value. Nevertheless, we'll include their definitions here for reference.

Part of the reason these inverse functions are so rarely used is that their domains and ranges are **disconnected**; that is, they consist of unions of connected intervals that are separated by discontinuities.

Definition: The fundamental domain of secant is

$$D = \left[0, \frac{\pi}{2}\right) \cup \left(\frac{\pi}{2}, \pi\right].$$

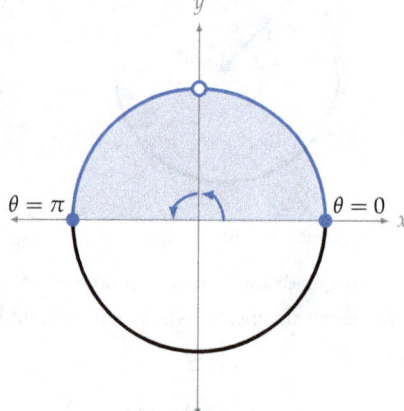

On this domain, the function $f(x) = \sec x$ has range

$$(-\infty, -1] \cup [1, \infty).$$

so we can define **the inverse secant function**:

$$f^{-1}: (-\infty, -1] \cup [1, \infty) \longrightarrow \left[0, \frac{\pi}{2}\right) \cup \left(\frac{\pi}{2}, \pi\right]$$

$$x \longmapsto \operatorname{arcsec} x$$

Definition: The fundamental domain of cosecant is

$$D = \left[-\frac{\pi}{2}, 0\right) \cup \left(0, \frac{\pi}{2}\right].$$

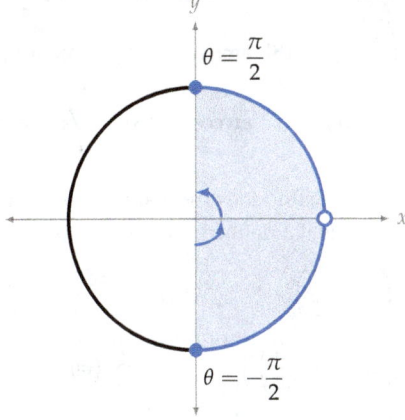

On this domain, the function $f(x) = \csc x$ has range

$$(-\infty, -1] \cup [1, \infty).$$

so we can define **the inverse cosecant function**:

$$f^{-1}: (-\infty, -1] \cup [1, \infty) \longrightarrow \left[-\frac{\pi}{2}, 0\right) \cup \left(0, \frac{\pi}{2}\right]$$

$$x \longmapsto \operatorname{arccsc} x$$

For secant, the issue is that the function is defined as the reciprocal of cosine:

$$\sec \theta = \frac{1}{\cos \theta}.$$

Secant is one-to-one over the same intervals as cosine, in particular on the interval $[0°, 180°]$. Secant is positive when cosine is positive, in Quadrant I, and negative when cosine is negative, in Quadrant II. But unlike cosine, secant has a **denominator** that equal to zero at $90°$, exactly in the center of its fundamental domain.

Aside from its domain, secant also has a range that is disconnected. Cosine has a range of values that lie in the interval $[-1, 1]$, and the values of secant are the reciprocals of these. Therefore the values of secant always have either

$$\sec \theta \leq -1 \quad \text{or} \quad 1 \leq \sec \theta.$$

Cosecant has the analogous problem, being defined as the reciprocal of sine:

$$\csc \theta = \frac{1}{\sin \theta}.$$

Cosecant has the same fundamental domain as sine, $[-90°, 90°]$, though we must exclude the point in the interior at which its denominator is equal to zero.

LESSON 7

EXERCISES

For Exercises 1–8, decide whether or not the given angle θ is in the fundamental domain of the given function. If it is, evaluate the function at the angle θ.

1. $\theta = 0$; $\sin(\theta)$
2. $\theta = \pi$; $\cos(\theta)$
3. $\theta = 180°$; $\tan(\theta)$
4. $\theta = -\frac{\pi}{2}$; $\sec(\theta)$
5. $\theta = 90°$; $\csc(\theta)$
6. $\theta = 0$; $\cot(\theta)$
7. $\theta = -30°$; $\sec(\theta)$
8. $\theta = \frac{5\pi}{6}$; $\sin(\theta)$

In Exercises 9–16, decide whether or not the given value x is in the domain of the given inverse function. If it is, evaluate the function at the value x.

9. $x = 0$; $\arcsin(x)$
10. $x = 2$; $\arccos(x)$
11. $x = \sqrt{3}$; $\arctan(x)$
12. $x = -\frac{\sqrt{3}}{2}$; $\arccos(x)$
13. $x = 0$; $\text{arccsc}(x)$
14. $x = -1$; $\text{arcsec}(x)$
15. $x = \frac{2\sqrt{3}}{3}$; $\text{arcsec}(x)$
16. $x = 0$; $\text{arccot}(x)$

Each expression below represents an angle θ. Determine the quadrant in which the terminal side of θ lies.

17. $\arcsin\left(\frac{\sqrt{3}}{2}\right)$
18. $\arctan\left(-\sqrt{3}\right)$
19. $\text{arcsec}(2)$
20. $\text{arccot}(-10.1)$
21. $\arccos(-0.8)$
22. $\text{arccsc}(5.7)$

For Exercises 23–28, find the exact value of each expression. If the expression is not defined, simply say "undefined."

23. $\cos\left(\sin^{-1}\frac{1}{2}\right)$
24. $\sin^{-1}\left(\cos\frac{5\pi}{6}\right)$
25. $\sec\left(\cos^{-1}\left(-\frac{3}{2}\right)\right)$
26. $\cos\left(\tan^{-1}0\right)$
27. $\cot^{-1}\left(\cos\left(-\pi\right)\right)$
28. $\cos^{-1}\left(\sin\left(-\frac{\pi}{4}\right)\right)$

Use a right triangle to write an algebraic expression that is equivalent to the given one.

29. $\sin\left(\arccos x\right)$
30. $\sin\left(\arctan x\right)$
31. $\cos\left(\arcsin 2x\right)$
32. $\tan\left(\arccos\frac{x}{5}\right)$
33. $\cot\left(\arctan x\right)$
34. $\csc\left(\arctan\frac{x}{\sqrt{2}}\right)$

LESSON 8

Solving Equations for Angles

The types of problems that we'll approach here are subtly different from what we saw in the last lesson, and broader, in a certain sense. The examples and exercises in Lesson 7 typically asked us to **evaluate** an arcfunction at a point, or to "find the exact value" of some expression. The wording of this, which asks for **the** value of an expression, implies that it should be unique.

But a more general question might ask something akin to "find **all** angles that satisfy the following equation," or possibly "find all solutions that lie in a given interval." These imply that there may be several such solutions, or none at all, and we'll have to consider how many solutions exist.

These two types of questions are not exclusive, by any means. On the one hand, finding all angles that satisfy an equation usually involves finding first the **principal angle** that makes the equation true, and then using it to find the rest of the solutions. This is particularly true for questions for which a calculator is needed.

On the other hand, we've seen that choosing the principal angle of an arcfunction basically amounts to selecting one specific angle from a list. Solving a more complex equation for an angle follows this same process, so the process of "evaluating an arcfunction" is equivalent to finding the solutions to an equation on the "smallest possible interval."

LESSON 8

Example: Find all solutions to the following equation that lie in the interval $[0, 2\pi]$:

$$2\sin\theta - \sqrt{2} = 0.$$

We first move the constants to one side of the equation to isolate the trigonometric function:

$$\sin\theta = \frac{\sqrt{2}}{2}.$$

Now we wish to find the angles θ that make this equation true. Applying the arcsine function to both sides and using a calculator will give the principal value

$$\theta = \arcsin\left(\frac{\sqrt{2}}{2}\right) = 45°.$$

In Lesson 7 we were asked to evaluate these types of **functions**, each of which requires a single, unique solution. But here we are asked to find a **set** of solutions to the equation the interval $[0, 2\pi]$. So we will look to the Unit Circle to find all points with the needed y-coordinate, not just those in the fundamental domain of sine. We find the two points P and Q labeled in the image at right. The former corresponds to the angle $\alpha = 45°$. The latter has the same reference angle, which means that $\beta = 180° - 45° = 135°$.

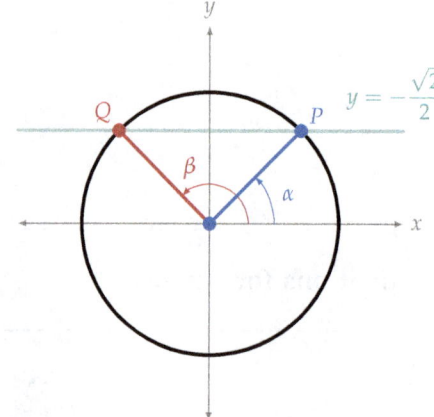

We conclude that these are the only **two** angles satisfying the equation on the interval $[0, 2\pi]$. Because these types of equations sometimes have a large number of solutions, we often write them as a set: $\theta \in \{45°, 135°\}$.

Example: Find all solutions to the equation:

$$2\cos\theta - \sqrt{3} = 0.$$

We again move the constants in this equation to one side, and look to find the angles θ that satisfy this equation.

84

The cosine function corresponds to the *x*-coordinate of the points on the Unit Circle, so we look for points having

$$x = \cos\theta = \frac{\sqrt{3}}{2}.$$

We could find these angles either by using a calculator or by recognizing the constant as the length of the long side of a 30/60/90 triangle. In either case there are **two** such points, labeled P and Q in the image below. They correspond, respectively, to the principal angle $\alpha = 30°$ and the non-principal angle $\beta = 330°$.

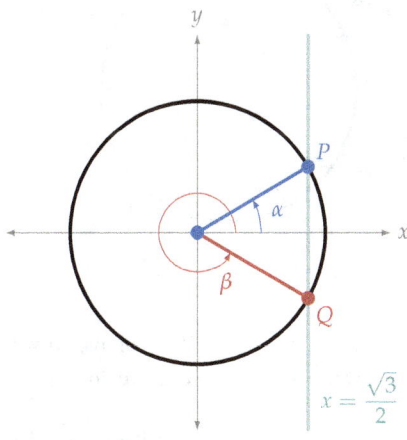

In fact, these points correspond to other, coterminal angles as well. The point P also corresponds to the angle $\alpha = 390°$, and again to the angle $\alpha = 750°$. We can write **every** angle that corresponds to P collectively, as a set:

$$\alpha \in \{\, 30° + 360°k \mid k \in \mathbb{Z} \,\}.$$

Here k is taken to be any **integer**, either positive or negative. Each choice of k gives a solution to our equation. Unlike the previous example, here we are asked to include **all** of these in our set of solutions.

Likewise all of the angles corresponding to the point Q may be written collectively as a set:

$$\beta \in \{\, 330° + 360°k \mid k \in \mathbb{Z} \,\}.$$

Each choice of k again yields a solution to our equation. We can compile all of these solutions from both α and β into a single set with the **union** symbol:

$$\theta \in \{\, 30° + 360°k \mid k \in \mathbb{Z} \,\} \cup \{\, 330° + 360°k \mid k \in \mathbb{Z} \,\}.$$

These two examples have been carefully chosen so they can be done without a calculator, if we know and recognize the trigonometric values coming from the standard triangles.

Lesson 8

Obviously, this will not always be the case. For general equations with random coefficients we will almost always need to to use a calculator to estimate the irrational angles that are solutions. However, most hand-held calculators return only the single principal value of an arc-function. We will infer any non-principal solutions ourselves.

Example: Use a calculator to find all the solutions:

$$4\sin\theta + 3 = 0.$$

While superficially similar to the previous example, after we move the constants to the right-hand side we find that

$$\sin\theta = -\frac{3}{4}.$$

As expected, this value does not correspond to any standard angle. We'll need to resort to a calculator. Calculators may return angles in either degrees or radians, depending on the settings, and in this case we find that

$$\theta = \arcsin(-0.75) \approx -0.848 \text{ rad}.$$

This is an estimate of the principal value of the arcsine function, which itself is an irrational number.

We note that it is negative, and in magnitude less than one radian, which places it in Quadrant IV of the x,y-plane. From the Unit Circle we see a second point having a y-coordinate of $y = -0.75$, lying in Quadrant III. This point, Q, corresponds to another angle of

$$\beta \approx (\pi + 0.848) \text{ rad} \approx 3.9896 \text{ rad}.$$

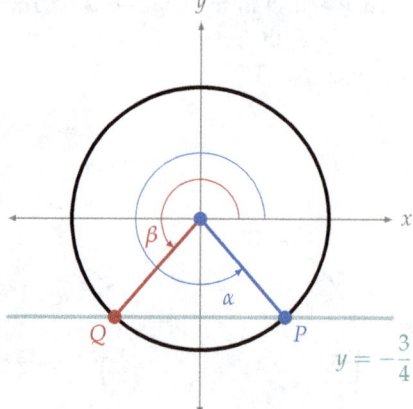

This example asks us to find **all** solutions, so we compile them with the same notation as our previous example:

$$\theta \in \{-0.848 + 2\pi k \mid k \in \mathbb{Z}\} \cup \{2.294 + 2\pi k \mid k \in \mathbb{Z}\}.$$

LESSON 8

We've just seen three examples that illustrate how to find and write the multitude of solutions that satisfy a given equation. Our next order of business is to do this when the argument of the trigonometric function is something more complex than a single angle. This is the case in our next example. Note also that we are asked to find solutions on a very specific interval.

Example: Find all solutions to the following equation that lie in the interval $[0, 4\pi]$:

$$\tan\left(\theta - \frac{\pi}{2}\right) = 1.$$

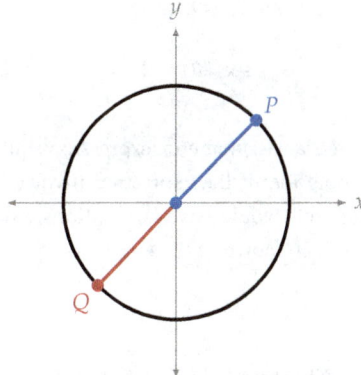

Any outside constants are already moved to the right-hand side, so we are ready to take the arctangent of both sides:

$$\theta - \frac{\pi}{2} = \arctan\left(\tan\left(\theta - \frac{\pi}{2}\right)\right) = \arctan(1).$$

There are **two** angles of the Unit Circle at which tangent is equal to 1, where sine and cosine are equal to each other. They are the principal angle $\alpha = 45°$ in Quadrant I, and the non-principal angle $\beta = 225°$ in Quadrant III:

$$\theta - \frac{\pi}{2} = \frac{\pi}{4} \qquad \theta - \frac{\pi}{2} = \frac{5\pi}{4}$$

$$\theta = \frac{3\pi}{4} \qquad \theta = \frac{7\pi}{4}$$

Of course, we should also add all coterminal angles to this set of solutions. We'll write all the solutions together as a union of two independent sets:

$$\theta \in \left\{\frac{3\pi}{4} + 2\pi k \mid k \in \mathbb{Z}\right\} \cup \left\{\frac{7\pi}{4} + 2\pi k \mid k \in \mathbb{Z}\right\}.$$

Now we are asked to find only those solutions that lie in the interval $[0, 4\pi]$. Plugging in different values for $k \in \mathbb{Z}$ will get us a variety of solutions. But exactly **four** of these angles will lie in the interval, corresponding to $k \in \{1, 2\}$:

$$\theta \in \left\{\frac{3\pi}{4}, \frac{7\pi}{4}, \frac{11\pi}{4}, \frac{15\pi}{4}\right\}.$$

87

LESSON 8

Example: Find all solutions to the equation

$$\sec(2\theta) = 2.$$

This example sets a trigonometric expression equal to 2. It's important to note that if the expression involved a sine or cosine function, this would have no solutions, since 2 is not in the range $[-1, 1]$. However, 2 **is** in the range of **secant**:

$$(-\infty, -1] \cup [1, \infty).$$

In fact, taking reciprocals of both sides of the equation gives

$$\cos(2\theta) = \frac{1}{2}.$$

This formulation may look a little more familiar to us, especially since 0.5 is the length of the short side of a 30/60/90 triangle. Therefore, with either a calculator or from our knowledge of the standard triangles, we find that

$$2\theta = \arccos\left(\cos\left(2\theta\right)\right) = \arccos\left(\frac{1}{2}\right) = 60°.$$

Since we are looking for all possible solutions, we should include all coterminal angles as well:

$$2\theta = 60° + 360°k.$$

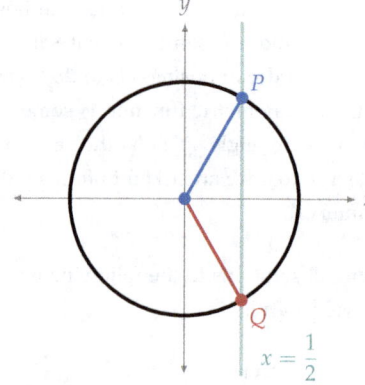

We solve this for θ by dividing everything in the equation by 2, **including** the added terms:

$$\theta = 30° + 180°k.$$

Likewise, we can find all angles corresponding to Q, using the non-principal angle $\beta = 300°$:

$$2\theta = 300° + 360°k$$

$$\theta = 150° + 180°k.$$

We write all of these solutions together in a single set:

$$\theta \in \{\, 30° + 180°k \mid k \in \mathbb{Z} \,\} \cup \{\, 150° + 180°k \mid k \in \mathbb{Z} \,\}.$$

LESSON 8

We're gradually increasing the complexity of these equations. The next step is to introduce some higher, second powers. For this we will borrow from the methods used to solve similar, non-trigonometric equations. In particular, one of the main methods will involve **factoring**.

The most important fact to keep in mind when factoring is the "Zero-Product Rule," which we list here:

Theorem: Suppose that a and b are two real numbers such that $ab = 0$. Then either $a = 0$ or $b = 0$.

In the context of solving equations, the terms a and b described above are typically expressions involving a variable. This will also be the case here.

Example: Find all solutions to the equation:
$$2\cos^2\theta + \cos\theta = 0.$$

To solve this, note that the expression $\cos\theta$ appears in both terms. Therefore we may think about factoring out this common term in the same way that we factor out the common factor of x in the (non-trigonometric) equation
$$2x^2 + x = 0.$$

The equivalence between these two equations can be made concrete by setting the two expressions equal to each other: $x = \cos\theta$. In any case, we would solve the latter equation by factoring out the common term:
$$x(2x+1) = 0.$$

We now have the two expressions, x and $2x+1$, which are multiplied to equal zero. By the theorem, either one or the other must itself be equal to zero. That is, either
$$x = 0 \quad \text{or} \quad 2x+1 = 0.$$

This leads to the two possible solutions
$$x = 0 \quad \text{or} \quad x = -\frac{1}{2}.$$

Now, returning to our context, we know that x represents the expression $x = \cos\theta$. So going through the same steps gets us to the two solutions
$$\cos\theta = 0 \quad \text{or} \quad \cos\theta = -\frac{1}{2}.$$

Now we have two equations to solve, which we'll do using the same methods as our earlier examples. Each equation will have its own infinite set of coterminal solutions.

LESSON 8

The solutions to the first equation, $\cos\theta = 0$, are shown in red below. The solutions to the second equation are shown in green. In the interval $[0°, 360°]$ we find that the equation has **four** solutions. Remember that here we do not need to distinguish between principal and non-principal angles.

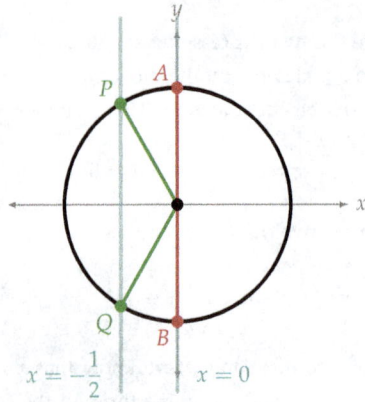

We first list these four solutions from the interval $[0°, 360°]$:

$$\theta \in \{90°, 120°, 240°, 270°\}.$$

Finally, we'll list all angles that are coterminal to these four. We could write these solutions together as a union of independent sets, though this would exceed our page margins.

As an alternative notation, we'll write them as a piecewise-defined equality. The angle θ may be equal to any of these four expressions, for any value of k:

$$\theta = \begin{cases} 90° + 360°k, & \text{for } k \in \mathbb{Z} \\ 120° + 360°k, & \text{for } k \in \mathbb{Z} \\ 240° + 360°k, & \text{for } k \in \mathbb{Z} \\ 270° + 360°k, & \text{for } k \in \mathbb{Z} \end{cases}$$

We'll also practice our factoring technique with an equation that has a constant term. Just as in the last example, we'll try to simplify the equation by using a temporary variable to stand in for our trigonometric expression.

Example: Find all solutions to the equation:

$$3\sin^2\theta + 4\sin\theta + 1 = 0$$

As we've just mentioned, we'll introduce a new variable to help us with the process of factoring. Here we set $u = \sin\theta$.

This transforms our equation into a non-trigonometric version, which we can factor into two parts:

$$3u^2 + 4u + 1 = (u+1)(3u+1).$$

Therefore our original equation could have been factored in the same way:

$$3\sin^2\theta + 4\sin\theta + 1 = 0$$

$$(\sin\theta + 1)(3\sin\theta + 1) = 0$$

By the Zero-Product Rule, we know that either of these two expressions must be equal to zero. This gives the solutions:

$$\sin\theta = -1 \quad \text{or} \quad \sin\theta = -\frac{1}{3}.$$

The solutions to the former equation are shown in red below, and those of the latter are shown in green. We see that this equation has **three** solutions in the interval $[0°, 360°]$.

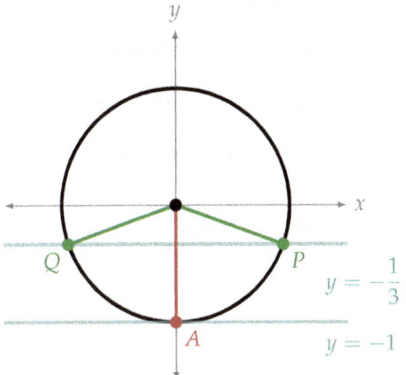

The solutions to the former equation are straightforward, and consist of the set of coterminal angles

$$\theta \in \{270° + 360°k \mid k \in \mathbb{Z}\}.$$

The other equation involves a non-standard value, so we'll need to use a calculator, which tells us that

$$\arcsin\left(-\frac{1}{3}\right) = -\arcsin\left(\frac{1}{3}\right) \approx -19.47°.$$

This is of course a principal value, which in the interval $[0°, 360°]$ corresponds to the angles $340.53°$ and $199.47°$. We'll list all of the angles that are coterminal to the three we've found, using the same piecewise notation as the last example. Note that the latter angles expressed by decimals are approximate values.

$$\theta = \begin{cases} 270° + 360°k, & \text{for } k \in \mathbb{Z} \\ 340.53° + 360°k, & \text{for } k \in \mathbb{Z} \\ 199.47° + 360°k, & \text{for } k \in \mathbb{Z} \end{cases}$$

Readers who are well versed in solving these types of non-trigonometric "quadratic" equations may feel free to use any standard method of finding solutions here. In a later lesson we will also review how to **complete the square**, which is very useful for equations that do not easily factor.

LESSON 8

EXERCISES

In Exercises 1–6, find all the solutions in the interval $[0, 2\pi]$.

1. $\cos\theta + 1 = 0$
2. $\tan x - \sqrt{3} = 0$
3. $\tan\theta + 1 = 0$
4. $\sec x - \sqrt{2} = 0$
5. $\sin^2 x - 1 = 0$
6. $\cos x = 3\cos x - 2$

For Exercises 7–10, solve each equation for x.

7. $2\arccos(x) = \pi$
8. $\arccos(x + 5) + \pi = 0$
9. $4\arctan(2x) = -\pi$
10. $7\arcsin(x) = 4\arcsin(x) - \pi$

For each equation below, identify the number of solutions that lie in the interval $[0, 2\pi]$.

11. $\sin\theta = 1$
12. $\sin\theta = \dfrac{1}{2}$
13. $\sin 2\theta = \dfrac{1}{2}$
14. $\sin 2\theta = \dfrac{3}{2}$
15. $\cos^2 3\theta = \dfrac{1}{2}$
16. $\tan\theta = -3$
17. $\cos 8\theta = 1$
18. $\sin\dfrac{1}{2}\theta = -\dfrac{\sqrt{3}}{2}$

Find all solutions that lie in the interval $[0, 2\pi]$.

19. $\sin x = \cos x$
20. $4\cos^2 x = 1$
21. $\cos(2\theta) = -\dfrac{1}{2}$
22. $\sin\left(3\theta + \dfrac{\pi}{18}\right) = 1$
23. $2\sin x \cos x = \sqrt{3}\sin x$
24. $\sin x = \tan x$
25. $2\sin^2 x + 1 = 3\sin x$
26. $2\sin x \cos x = \sqrt{2}\cos x$

For Exercises 27–32, give a general formula that describes all solutions to the equation.

27. $2\sin\theta = \sqrt{3}$
28. $\tan\theta = -1$
29. $4\sec\theta + 6 = -2$
30. $2\cos^2(2x) + \cos(2x) = 1$
31. $2\sin x \cos x - \sin x = 0$
32. $\tan^2 x + \tan x = 0$

Use a calculator to find the solutions to the equation in the interval $[0, 2\pi]$. Round answers to two decimal places.

33. $\cos\theta = -0.9$
34. $5\tan\theta + 9 = 0$
35. $15\sin\theta = 2$
36. $4\cos\theta - 3 = 0$

LESSON 9

Using Trigonometric Identities

Equations that involve trigonometric functions can reach levels of complexity that rival any other algebraic equation. Solving such equations typically requires us to blend the strategies of general algebra, like factoring, together with our knowledge of the trigonometric functions and their definitions. More complicated equations will naturally require larger resources from both of these areas.

To help, it's considered essential to have a working library that describes the variety of relationships that exist between trigonometric functions. We've seen some of these relationships before in Lessons 3 and 5, and we may have even used some of them already to solve some simple equations.

But of course, there are so many more of these identities that we would have a hard time describing all of them in a single lesson. Instead we'll concentrate on a small collection of identities, which are all derivatives of a very special pair known as the **Sum and Difference Formulas**.

For the new identities that we're seeing for the first time, we'll spend a good part of this lesson describing how the relationships arise. We'll also see several examples that use the identities to perform direct calculations. However, our end goal is to be able to use these identities to solve equations in the same way that we saw in Lesson 8. We should not lose sight of that fact here.

LESSON 9

THE SUM/DIFFERENCE FORMULAS

We've noted before, in Lesson 6 specifically, that for nearly all functions f the expressions $f(x+y)$ and $f(x)+f(y)$ are not equal. This is especially true in the case of the trigonometric functions. Here we'll determine the correct formulas for sums and differences of angles.

At right we have an image consisting of a rectangle that has been split into four different triangles, all of which are right triangles. We'll first focus on the many **angles** in the image.

The yellow triangle on the bottom of the image has an angle marked as α, so the opposing angle measures $90° - \alpha$.

This opposing angle, together with the blue right angle and the smaller angle from the yellow triangle on the top, form a straight line. Therefore their sum is $180°$, and we can then infer that this third angle is again equal to α:

$$(90° - \alpha) + (90°) + (\alpha) = 180°.$$

Thus the two yellow triangles are **similar**.

Next, in the bottom left corner of the image, we see that the missing purple angle measures $90° - \alpha - \beta$. So the opposing purple angle at the top of the image must be equal to $\alpha + \beta$.

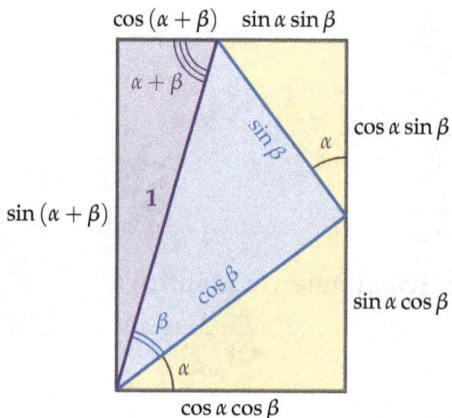

Now that all of the angles have been determined, we focus on the blue triangle. This has a hypotenuse of length one, so the sides adjacent and opposite to β have lengths of $\cos \beta$ and $\sin \beta$, respectively.

For the same reason, the adjacent and opposite sides of the purple angle $\alpha + \beta$ have lengths $\cos(\alpha + \beta)$ and $\sin(\alpha + \beta)$.

Finally, we return to the yellow triangles. The lower triangle has a hypotenuse of length $\cos \beta$, and the side that is adjacent to α also forms the base of the greater rectangle.

LESSON 9

If we label the length of this adjacent side as x, then

$$\cos \alpha = \frac{x}{\cos \beta},$$

which can be solved to give

$$x = \cos \alpha \cos \beta.$$

Likewise, the opposite side has a length of $\sin \alpha \cos \beta$.

The upper yellow triangle has a hypotenuse of length $\sin \beta$. With the same argument, the sides adjacent and opposite to the angle α, respectively, have lengths

$$\cos \alpha \sin \beta,$$

$$\sin \alpha \sin \beta.$$

Afterward, we simply compare the lengths of the opposing sides of the full rectangle. Since the two vertical edges are the same length, we find that

$$\sin(\alpha + \beta) = \sin \alpha \cos \beta + \cos \alpha \sin \beta.$$

We write the horizontal edges as a difference:

$$\cos(\alpha + \beta) = \cos \alpha \cos \beta - \sin \alpha \sin \beta.$$

These are our formulas for the sine and cosine of a sum. Unfortunately, our drawing works only for small angles α and β, but this drawing can be modified to verify the formulas for larger angles.

Also, by relabeling the angles and following the same arguments we could obtain the formulas for **differences**:

$$\sin(\alpha - \beta) = \sin \alpha \cos \beta - \cos \alpha \sin \beta,$$

$$\cos(\alpha - \beta) = \cos \alpha \cos \beta + \sin \alpha \sin \beta.$$

This set of formulas expands the number of angles that we can now calculate exactly, as the following example shows.

Example: Find the exact value of $\sin(75°)$.

We simply write $75°$ as a sum of standard angles, and apply the appropriate formula:

$$\sin(75°) = \sin(30° + 45°)$$
$$= \sin(30°)\cos(45°) + \cos(30°)\sin(45°)$$
$$= \left(\frac{1}{2}\right)\left(\frac{\sqrt{2}}{2}\right) + \left(\frac{\sqrt{3}}{2}\right)\left(\frac{\sqrt{2}}{2}\right)$$
$$= \frac{\sqrt{2} + \sqrt{6}}{4}.$$

Lesson 9

Every time that we have a formula for sine and cosine, as we do here, we inherit similar formulas for each of the other trigonometric functions. For example, for the tangent of a sum we can write

$$\tan(\alpha + \beta) = \frac{\sin(\alpha + \beta)}{\cos(\alpha + \beta)} = \frac{\sin\alpha\cos\beta + \cos\alpha\sin\beta}{\cos\alpha\cos\beta - \sin\alpha\sin\beta}.$$

To put everything in terms of tangent, we divide both the numerator and denominator by $\cos\alpha\cos\beta$:

$$\tan(\alpha + \beta) = \frac{\sin\alpha\cos\beta + \cos\alpha\sin\beta}{\cos\alpha\cos\beta - \sin\alpha\sin\beta} \cdot \frac{\frac{1}{\cos\alpha\cos\beta}}{\frac{1}{\cos\alpha\cos\beta}}$$

$$\tan(\alpha + \beta) = \frac{\frac{\sin\alpha}{\cos\alpha} + \frac{\sin\beta}{\cos\beta}}{1 - \frac{\sin\alpha\sin\beta}{\cos\alpha\cos\beta}}$$

$$\tan(\alpha + \beta) = \frac{\tan\alpha + \tan\beta}{1 - \tan\alpha\tan\beta}$$

There is a similar formula for the tangent of a difference:

$$\tan(\alpha - \beta) = \frac{\tan\alpha - \tan\beta}{1 + \tan\alpha\tan\beta}.$$

These formulas are used to evaluate trigonometric functions at any angle that is a sum or difference of the standard angles, such as $105°$, which can be written as $135° - 30°$.

Example: Find the exact value of

$$\sin\left[\arcsin\left(\frac{8}{17}\right) - \arccos\left(-\frac{5}{13}\right)\right].$$

The two arcfunctions in this expression each represent an angle, so it would be appropriate to give them names:

$$\alpha = \arcsin\left(\frac{8}{17}\right), \qquad \beta = \arccos\left(-\frac{5}{13}\right).$$

Thus the expression can be rewritten $\sin(\alpha - \beta)$, leading to the Difference Formula for sine:

$$\sin(\alpha - \beta) = \sin\alpha\cos\beta - \cos\alpha\sin\beta$$

We're now left to find the sines and cosines of α and β. We could do this by estimating these irrational angles and then further estimating their sines and cosines, but we will do something more exact.

We've drawn each of the angles in the images at right. Since

$$\sin\alpha = \frac{8}{17},$$

the triangle that is formed has an opposite side of length 8 and a hypotenuse of length 17. Because both values are positive, it lies in the first quadrant.

The reference triangle for β is formed in the same way, but because $\cos\beta$ is negative, it lies in the second quadrant.

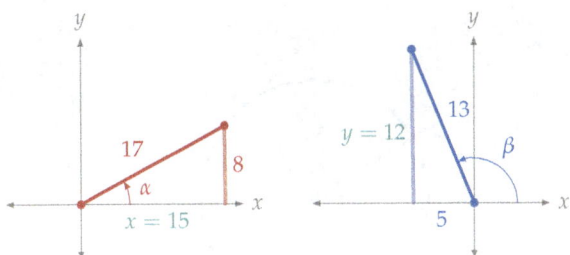

Once the triangles are drawn with the information we know, we fill in the missing sides with the Pythagorean Theorem:

$$x = \sqrt{17^2 - 8^2} = \sqrt{289 - 64} = \sqrt{225} = 15,$$

$$y = \sqrt{13^2 - 5^2} = \sqrt{169 - 25} = \sqrt{144} = 12.$$

With all three sides determined for each triangle, we can find any trigonometric values that we like. In particular,

$$\sin\alpha = \frac{8}{17} \qquad \sin\beta = \frac{12}{13}$$

$$\cos\alpha = \frac{15}{17} \qquad \cos\beta = -\frac{5}{13}$$

At this point we can fill in all of the blanks in the Difference Formula and simplify:

$$\sin\left[\arcsin\left(\frac{8}{17}\right) - \arccos\left(-\frac{5}{13}\right)\right] = \sin[\alpha - \beta]$$

$$= \sin\alpha\cos\beta - \cos\alpha\sin\beta$$

$$= \left(\frac{8}{17}\right)\left(-\frac{5}{13}\right) - \left(\frac{15}{17}\right)\left(\frac{12}{13}\right)$$

$$= -\frac{40}{221} - \frac{180}{221}$$

$$= -\frac{220}{221}.$$

All of the Sum and Difference Formulas for sine and cosine use these same four expressions, in different combinations. Therefore most of these problems will proceed exactly as this one has. The only difference will occur at the end when we select the particular formula needed, which indicates the signs and the arrangement of the terms.

Aside from using the Sum and Difference Formulas directly to evaluate trigonometric functions, we can also use them to solve trigonometric equations. What follows is a non-obvious example, where a seemingly innocuous equation requires a sum formula to solve.

LESSON 9

Example: Find all solutions to the following equation:

$$\sqrt{3}\sin\theta + \cos\theta = 1$$

Any attempt to solve this equation for θ is likely to be fruitless. Trying to apply arcsine or arccosine to both side will not work, because the arc-functions are not additive any more than the trigonometric functions are, and there will be no way to simplify the expression that results.

However, we may recognize the coefficients of this equation as coming from a 30/60/90 triangle, or at least, we could set up a 30/60/90 triangle with those side lengths. In fact, if we divide the equation by 2, the coefficients will be exactly the standard values we know:

$$\frac{\sqrt{3}}{2}\sin\theta + \frac{1}{2}\cos\theta = \frac{1}{2}.$$

We could use either of the angles 30° or 60° for the next operation. Since $\cos 30° = \sqrt{3}/2$ and $\sin 30° = 1/2$, we substitute those values and then apply the sum formula:

$$(\cos 30°)\sin\theta + (\sin 30°)\cos\theta = \frac{1}{2}$$

$$\sin(\theta + 30°) = \frac{1}{2}.$$

We're now in a position to solve this equation with the methods that we learned in Lesson 8.

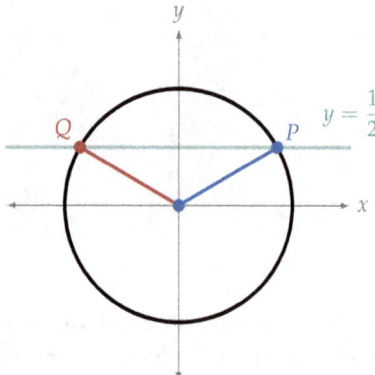

In the Unit Circle the sine function is equal to 0.5 at 30° and 150°, so we obtain the equations

$$\theta + 30° = 30° \qquad \theta + 30° = 150°$$
$$\theta = 0° \qquad \theta = 120°$$

Collecting these with all of their coterminal angles in a single set gives us the solutions

$$\theta \in \{360°k \mid k \in \mathbb{Z}\} \cup \{120° + 360°k \mid k \in \mathbb{Z}\}.$$

To the careful reader this last example may have seemed a bit contrived. After all, the coefficients of the given equation were almost exactly the values of a 30/60/90 triangle. But can we solve an equation of this type in which the coefficients are random numbers? The answer is yes, though we will most likely need to use a calculator along the way.

Example: Find all solutions to the following equation:

$$3\cos\theta + 5\sin\theta = 2$$

We'll try to use the same strategy as the last example, in which we write the coefficients as the sine and cosine of an angle and then apply the sum formula. That formula is witten here in terms of the angles ϕ and θ:

$$\sin(\phi + \theta) = \sin\phi\cos\theta + \cos\phi\sin\theta$$

It doesn't particularly matter which of the four sum formulas is used here. However, this formula is sometimes preferred because everything commutes, and we don't have to worry about the order in which anything is written.

We see that the coefficient 3 will be associated to $\sin\phi$, and 5 with $\cos\phi$, so we first draw a right triangle with opposite and adjacent sides of these lengths, respectively.

The length of the missing hypotenuse is simple to find with the Pythagorean Theorem, and it is filled in here.

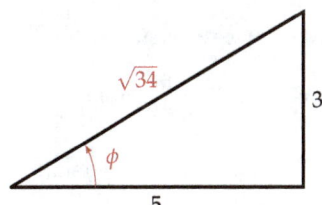

Now we can estimate ϕ using a calculator,

$$\phi = \arctan\left(\frac{3}{5}\right) \approx 30.96°,$$

and also be more exact about the sine and cosine of ϕ:

$$\sin\phi = \frac{3}{\sqrt{34}}, \qquad \cos\phi = \frac{5}{\sqrt{34}}.$$

These are not quite the coefficients that we saw the original equation. However, we'll set up a direct substitution by first dividing that equation by the length of the hypotenuse:

$$\frac{3}{\sqrt{34}}\cos\theta + \frac{5}{\sqrt{34}}\sin\theta = \frac{2}{\sqrt{34}}.$$

LESSON 9

After replacing these coefficients with the sine and cosine expressions, we apply the Sum Formula for sine. All of the steps so far are shown below.

$$3\cos\theta + 5\sin\theta = 2$$

$$\frac{3}{\sqrt{34}}\cos\theta + \frac{5}{\sqrt{34}}\sin\theta = \frac{2}{\sqrt{34}}$$

$$\sin(30.96°)\cos(\theta) + \cos(30.96°)\sin(\theta) = \frac{2}{\sqrt{34}}$$

$$\sin(30.96° + \theta) = \frac{2}{\sqrt{34}}$$

$$30.96° + \theta = \arcsin\left(\frac{2}{\sqrt{34}}\right).$$

A calculator shows that the right-hand side is about

$$\arcsin\left(\frac{2}{\sqrt{34}}\right) \approx 20.06°.$$

This is the principal angle. We find the non-principal angle by taking a quick look at the Unit Circle. Sine is positive in Quadrants I and II, and the angle in Quadrant II with reference angle 20.06° is

$$180° - 20.06° = 159.94°.$$

These are the two angles we'll use in our equation.

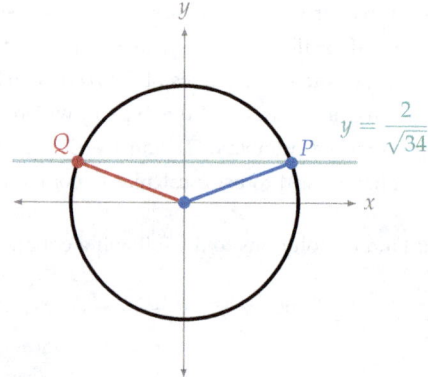

Starting with the principal angle, the approximation

$$\theta + 30.96° \approx 20.06°$$

gives an estimate of $\theta \approx -10.9°$ for the solution, which is coterminal to $\theta \approx 349.1°$. Likewise, the approximation

$$\theta + 30.96° \approx 159.94°$$

gives an estimate of $\theta \approx 128.98°$. The collection of all angles coterminal to these is written below as a union of two sets.

$$\theta \in \{349.1° + 360°k \mid k \in \mathbb{Z}\} \cup \{128.98° + 360°k \mid k \in \mathbb{Z}\}.$$

LESSON 9

THE DOUBLE-ANGLE FORMULAS

The formulas written below are really nothing more than the Sum Formulas we learned earlier, in the case that $\alpha = \beta$.

$$\sin(2\alpha) = 2 \sin \alpha \cos \alpha$$

$$\cos(2\alpha) = \cos^2 \alpha - \sin^2 \alpha$$

$$\tan(2\alpha) = \frac{2 \tan \alpha}{1 - \tan^2 \alpha}$$

These common forms are sometimes arranged differently. For example, applying either of the equalities

$$\cos^2 \alpha = 1 - \sin^2 \alpha \quad \text{or} \quad \sin^2 \alpha = 1 - \cos^2 \alpha$$

to the second equation will give us the following identities:

$$\cos(2\alpha) = 1 - 2 \sin^2 \alpha,$$

$$\cos(2\alpha) = 2 \cos^2 \alpha - 1.$$

The five of these formulas are collectively known as the Double-Angle Formulas. They can be used to directly evaluate trigonometric functions of particular angles, in exactly the same way as the Sum Formulas. For our example here we will use the formulas to solve an equation.

Example: Find all solutions to the equation that lie in the interval $[0°, 360°]$:

$$3 \sin \theta + \cos 2\theta = 2.$$

The argument of the cosine function is doubled here, which gives us three choices of Double-Angle Formula to use. To put everything in terms of a single function (sine), we make the following substitution and simplify:

$$3 \sin \theta + \cos(2\theta) = 2$$

$$3 \sin \theta + (1 - 2 \sin^2 \theta) = 2$$

$$3 \sin \theta - 2 \sin^2 \theta - 1 = 0$$

$$2 \sin^2 \theta - 3 \sin \theta + 1 = 0$$

$$(2 \sin \theta - 1)(\sin \theta - 1) = 0.$$

This leaves us with the two equations

$$\sin \theta = \frac{1}{2} \quad \text{or} \quad \sin \theta = 1$$

The solutions to the first equation that lie in the interval $[0°, 360°]$ are $\theta = 30°$ and $\theta = 150°$. The only solution to the second is $\theta = 90°$. All three solutions are written below.

$$\theta \in \{30°, 90°, 150°\}.$$

LESSON 9

THE HALF-ANGLE FORMULAS

For the next set of formulas we'll start with two of the Double-Angle Formulas that we just learned:

$$\cos(2\alpha) = 1 - 2\sin^2\alpha,$$
$$\cos(2\alpha) = 2\cos^2\alpha - 1.$$

Solving each of these for the squared terms gives

$$\sin^2\alpha = \frac{1}{2}(1 - \cos 2\alpha),$$
$$\cos^2\alpha = \frac{1}{2}(1 + \cos 2\alpha).$$

Note that the angle on the left side of each equation is **half** of the angle on the right. We can emphasize this fact by rewriting the formulas:

$$\sin^2\left(\frac{\alpha}{2}\right) = \frac{1}{2}(1 - \cos\alpha)$$
$$\cos^2\left(\frac{\alpha}{2}\right) = \frac{1}{2}(1 + \cos\alpha)$$

In either form, these are known as the Half-Angle Formulas. Sometimes these are further solved for the half-angled sides by taking the square root of both sides. However, this can be problematic since there are both positive and negative square roots, and it is often not clear which of the two we should choose. We'll show this version of the formulas for completeness and then look at an example below.

$$\sin\left(\frac{\alpha}{2}\right) = \pm\sqrt{\frac{1-\cos\alpha}{2}} \qquad \cos\left(\frac{\alpha}{2}\right) = \pm\sqrt{\frac{1+\cos\alpha}{2}}$$

Example: Find the exact value of $\cos(157.5°)$.

The angle $157.5°$ is half of $315°$, which is a standard angle that we can evaluate exactly. However, because $157.5°$ lies in Quadrant II, where the cosine function has negative values, we expect $\cos(157.5°)$ to be a **negative** number.

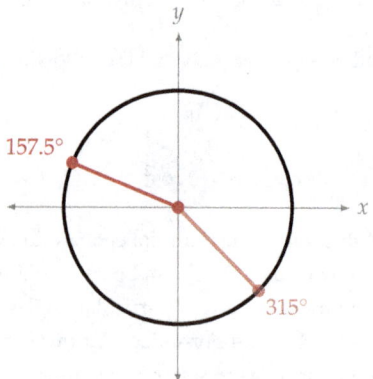

We'll use a Half-Angle Formula for the calculation:

$$\cos^2(157.5°) = \frac{1}{2}(1 + \cos 315°)$$

$$= \frac{1}{2}\left(1 + \frac{\sqrt{2}}{2}\right)$$

$$= \frac{1}{2}\left(\frac{2 + \sqrt{2}}{2}\right)$$

$$\cos^2(157.5°) = \frac{2 + \sqrt{2}}{4}.$$

We next take the square root of both sides. Note that we choose the **negative** square root.

$$\cos(157.5°) = -\sqrt{\frac{2 + \sqrt{2}}{4}} = -\frac{1}{2}\sqrt{2 + \sqrt{2}}.$$

Example: Find the exact value of $\sin\left(\frac{7\pi}{12}\right)$.

This time doubling the angle $7\pi/12$ gives us $7\pi/6$, and:

$$\sin\left(\frac{7\pi}{6}\right) = -\frac{\sqrt{3}}{2}.$$

However, $7\pi/12$ lies in Quadrant II, so we should expect our solution to be a **positive** number.

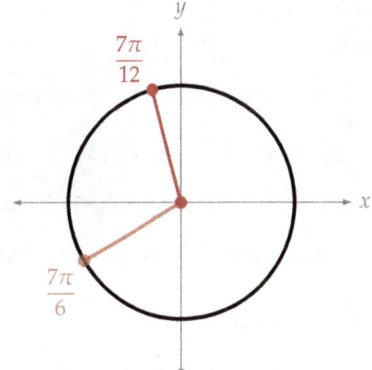

Using a Half-Angle Formula gives us

$$\sin^2\left(\frac{7\pi}{12}\right) = \frac{1}{2}\left(1 - \cos\left(\frac{7\pi}{6}\right)\right)$$

$$= \frac{1}{2}\left(1 - \left(-\frac{\sqrt{3}}{2}\right)\right)$$

$$\sin^2\left(\frac{7\pi}{12}\right) = \frac{2 + \sqrt{3}}{4}.$$

Finally, we choose the **positive** square root below.

$$\sin\left(\frac{7\pi}{12}\right) = \sqrt{\frac{2 + \sqrt{3}}{4}} = \frac{1}{2}\sqrt{2 + \sqrt{3}}.$$

LESSON 9

EXERCISES

For Exercises 1–12, find the exact value of each expression.

1. $\sin 165°$
2. $\cos 75°$
3. $\sin 285°$
4. $\tan 105°$
5. $\cos 15°$
6. $\tan(-15°)$
7. $\cos 345°$
8. $\tan 195°$
9. $\sin 255°$
10. $\cos 165°$
11. $\sin 67.5°$
12. $\cos 105°$

Find the exact value of each expression.

13. $\cos 25° \cos 5° - \sin 25° \sin 5°$
14. $\sin 40° \cos 20° + \cos 40° \sin 20°$
15. $\cos 80° \cos 20° + \sin 80° \sin 20°$
16. $\sin 65° \cos 35° - \cos 65° \sin 35°$

For Exercises 17–22, use trigonometric identities to find the exact value of each expression.

17. $\cos\left(\cos^{-1} 0 + \sin^{-1} \frac{1}{2}\right)$
18. $\tan\left(2\cos^{-1}\left(-\frac{3}{5}\right)\right)$
19. $\sin^2\left(\frac{1}{2}\cos^{-1} \frac{3}{5}\right)$
20. $\sin\left(\cos^{-1} \frac{5}{13} - \cos^{-1} \frac{4}{5}\right)$
21. $\sin\left(\tan^{-1} \sqrt{3} - \sin^{-1} \frac{1}{2}\right)$
22. $\cos\left(2\sin^{-1} \frac{3}{5}\right)$

For Exercises 23–28, use the figures below to find the exact value of each expression.

23. $\sin(\alpha + \beta)$
24. $\cos\left(\frac{\alpha}{2}\right)$
25. $\tan(\alpha - \beta)$
26. $\sin(2\beta)$
27. $\cos(\alpha + \beta)$
28. $\tan(2\alpha)$

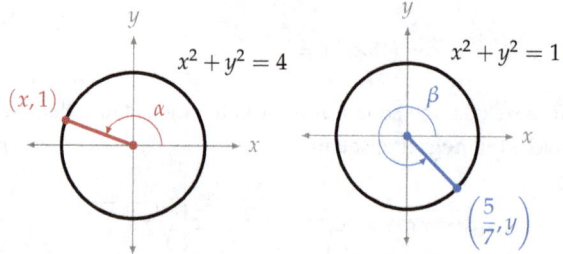

Find all solutions that lie in the interval $[0, 2\pi]$.

29. $\sin 2x = 2 \cos x$
30. $\sin^2 x = \cos^2 x - 1$
31. $3\cos 2x - 5\cos x = 1$
32. $\cos x = 1 - \sin x$

33. Find all solutions to the equation: $\sin \theta - \sqrt{3} \cos \theta = 1$.

34. Find all solutions to the equation: $\cot \theta + \csc \theta = -\sqrt{3}$.

LESSON 10

Applications of Trigonometry

From the multitude of problems that we've seen so far, it should be clear that trigonometry is useful in a wide variety of real-world problems. In this section we'll be focusing on these types of "word problems," and we'll try to be at least somewhat systematic in our approach.

Along the way we'll learn a bit of vocabulary that comes along with trigonometric problems in actual practice. It is good to note that nearly all solutions will be approximate here. This is normal when working with real data.

Though every problem is different, we'll adopt a few strategies to help us organize the data that are given to us.

- Identify all numerical values and the physical attributes that they describe. Assign a variable to each.

- Identify which attribute you are trying to find and assign that a variable as well.

- Draw a picture of the relationship between the variables. Establish a coordinate system over your picture. This picture will usually involve one or more triangles.

- From your picture, find an equation that relates your variables. Substitute in the values that you know and solve the equation.

LESSON 10

Example: The string of a kite, when being pulled tightly, makes an angle of 70° with the flat ground. If the string is 65 meters long, how high is the kite above the ground?

According to the steps we just listed, we should first identify all numerical values and assign variables to them.

- $\theta = 70°$ (angle between string and ground)
- $l = 65\,\text{m}$ (length of string)
- $(x, y) = ?$ (position of the kite)

The question only asks for the **height** of the kite, not its horizontal position. But it will not hurt us to assign an extra variable to the x-coordinate of the kite as we set up our coordinate system.

We'll establish this coordinate system so that the origin is at the base of the string, with the person who is holding the kite. As usual, the vertical direction is labeled as y and the horizontal as x. Furthermore, we consider the positive values of y to be measured in the upward direction. Since the distances in our problem are given in meters, this will be the unit of measure in our coordinate system. The image at right displays the situation in question.

In the image we find a right triangle that may be used as a reference triangle. Namely, the hypotenuse has length $l = 65\,\text{m}$, and the side opposite the angle $\theta = 70°$ has a length of y.

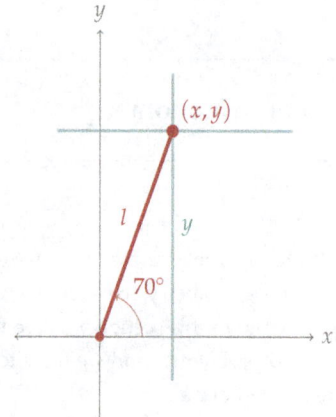

We therefore set up the equation

$$\sin 70° = \frac{y}{65\,\text{m}},$$

and solve for y to find the height of the kite:

$$y = (65\,\text{m})(\sin 70°) \approx (65)(0.93969)\,\text{m} \approx 61.08\,\text{m}.$$

Example: A pendulum is hanging from a string that is 50 centimeters long. Keeping the string tight, the plumb bob is moved 30° from its resting position. Find the exact vertical distance that the tip of the pendulum rises.

To approach this problem, we first highlight all the numerical values and assign them variables.

- $l = 50$ cm   (length of string)
- $\theta = 30°$ (angle the pendulum is moved)
- $(x, y) = ?$ (new position of the tip of the pendulum)

Note that we are actually asked for the distance that the pendulum **rises**, which is a difference of two heights. However, once we've found the new position (x, y), it will be easy to find this difference by subtracting the final y-value from the initial one.

Next we'll draw two pictures that illustrate the situation. The first shows the unmoved pendulum at rest, and the second shows the pendulum after it has been moved 30°. We establish a coordinate system as we draw the images. Since the top of the pendulum remains fixed through this movement, it makes sense to place the origin there.

All units in our system are measured in the centimeters that are indicated in the problem. If we label the vertical axis as y, and take the positive values to be pointing up, then the resting position of the pendulum will be at $(0, -50)$.

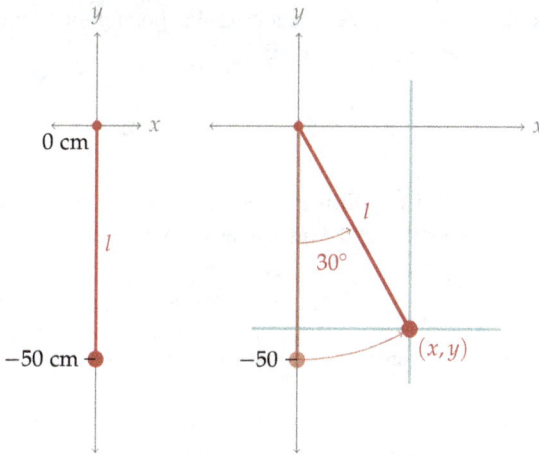

After moving the pendulum through the angle $\theta = 30°$, as indicated in the second picture, we see a right triangle emerge. This is a 30/60/90 triangle, which we can use as a reference triangle, with an adjacent side of length y, an opposite side of length x, and a hypotenuse of length $l = 50$.

LESSON 10

The reference triangle gives us the equations:

$$\cos 30° = \frac{y}{50\,\text{cm}},$$
$$\sin 30° = \frac{x}{50\,\text{cm}}.$$

Taking into account the locations of this point in our coordinate system, we conclude that

$$(x,y) = \left(50 \cdot \frac{1}{2}, -50 \cdot \frac{\sqrt{3}}{2}\right) = \left(25, -25\sqrt{3}\right).$$

We were asked to find the change in height of the plumb bob, so we'll subtract the final y-value from the initial one:

$$(-25\sqrt{3}) - (-50) = 50 - 25\sqrt{3}.$$

This is an exact answer, approximately equal to 6.7 cm.

The next example is possibly more straightforward than the last, but it explores a definition that will be the focus of many exercises in this lesson.

Example: A hiker is looking at a radio tower that is 300 meters tall. The angle between the flat ground and her line of sight to the top of the tower is 11°. How far is she from the base of the tower?

We first assign variable names to all of the numerical values.

- $(x,y) = (0, 300)$ (position of the top of the tower)
- $\theta = 11°$ (angle between ground and top of tower)
- $x = ?$ (position of the hiker)

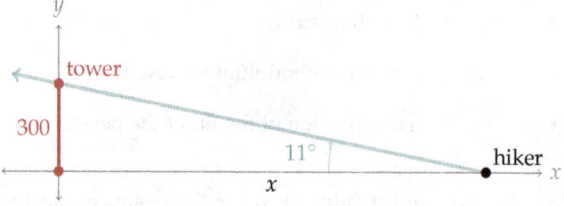

We see a right triangle formed in this image, with an opposite side of length 300 meters. We wish to find the length of the adjacent side. Therefore we use the tangent function:

$$\tan 11° = \frac{300\,\text{m}}{x}.$$

Solving this for x gives us the approximate distance between the hiker and the base of the tower:

$$x = \frac{300\,\text{m}}{\tan 11°} \approx \frac{300}{0.194}\,\text{m} \approx 1543.366\,\text{m}.$$

LESSON 10

ANGLES OF ELEVATION AND DEPRESSION

Imagine that a man is on a boat on a calm lake, so that the surface of the water is relatively level. The man looks up and sees a passing airplane. From the man's point of reference, the angle formed by the horizontal surface of the water and his line of sight to the plane is called the **angle of elevation** of the airplane.

Conversely, suppose that a woman piloting the airplane looks down and sees the boat on the surface of the lake. The angle formed by her line of sight to the boat and the horizontal is called the **angle of depression**.

Assuming that both observers make their measurements at the same time, or that nothing about their positions has changed in between measurements, then the angles of elevation and depression will have equal measures.

Example: In a parking garage, each floor is 20 feet apart. The ramp between two floors is 120 feet long. What is the angle of elevation of the ramp?

For this problem we'll stop short of setting up a full coordinate system, since a simple image of two floors and the ramp between them should be sufficient for our purposes.

We identify all of the numerical values given; the unknown variable here is the angle of elevation θ.

- $y = 20\,\text{ft}$ (height between floors)
- $r = 120\,\text{ft}$ (length of the ramp)
- $\theta = ?$ (angle of elevation)

An image of two floors and the ramp connecting them helps us visualize the problem, and how the variables fit together.

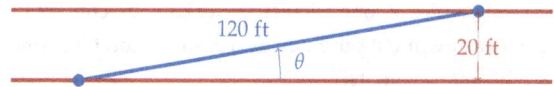

We see a right triangle formed by the height $y = 20\,\text{ft}$, the hypotenuse $r = 120\,\text{ft}$, and the unknown angle θ. The three are related by the sine function:

$$\sin\theta = \frac{20\,\text{ft}}{120\,\text{ft}} = \frac{1}{6}.$$

Now we can apply the arcsine function to both sides of the equation. A calculator can be used to approximate θ:

$$\theta = \arcsin(1/6) \approx 9.594°.$$

109

LESSON 10

Example: When the sun has an elevation of 31°20′ with the horizon, a flagpole casts a shadow that is 40 feet long. How tall is the flagpole?

For our variables, we can make the following assignments:

- $\theta = 31°20'$ (angle of elevation of the sun)
- $s = 40\,\text{ft}$ (length of the shadow)
- $(x, y) = ?$ (position of the top of the flagpole)

To make it easier on our calculators, we can also convert the angle measure into a pure degree measurement. Of course, this will be an estimate:

$$31°20' = 31° + \left(\frac{1}{3}\right)° \approx (31.333)°.$$

We have several options as we establish a coordinate system for this situation. In the image we've placed our origin at the tip of the shadow, which lies 40 feet away from the base of the flagpole. This has the advantage of keeping all of our values in the first quadrant, where everything is positive. Another good choice is to place the origin at the base of the flagpole. However, in that case the sun and shadow would lie in opposing quadrants.

Once our coordinate system is established and we draw all of the known values into it, we see another right triangle emerge. The side adjacent to θ has length 40 feet, and the side opposite has length y.

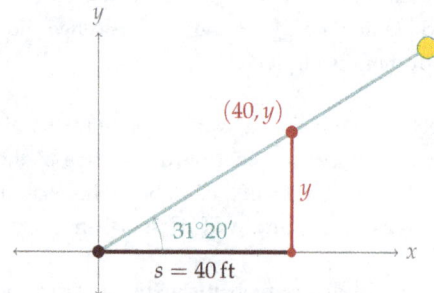

Using the tangent function, we have the equation

$$\tan(31.333°) \approx \frac{y}{40\,\text{ft}}.$$

Solving this for y gives us

$$y \approx (40\,\text{ft})(\tan 31.333°)$$
$$\approx (40)(0.6088)\,\text{ft}$$
$$\approx 24.352\,\text{ft}.$$

LESSON 10

NAVIGATIONAL BEARINGS

The definitions that follow have historically been associated with navigation at sea. However, many nautical conventions and terminology were later adopted by the aerospace industry. The terms here are no exception.

When a ship is out to open sea, all direction and orientation is typically measured with respect to the **forward** part of the ship, which points in the direction of the ship's motion. To an observer facing the direction of this motion, the ship has two sides. The **starboard** side lies to the observer's right, while the **port** side lies to their left.

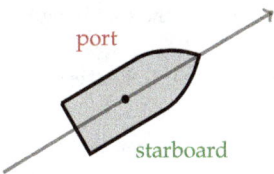

These two sides are not only distinguished by people on the ship itself, but also by observers on land or on a different ship entirely. Modern vessels equip colored lights on each side, with the starboard side showing a green light, and the port side showing red. At night, the lights allow the ship to be seen. The colors indicate how the ship is oriented, so observers can infer the direction that it is traveling.

Modern airplanes also equip red and green lights for the same reasons. Typically these are carried on the tips of the airplane's wings.

In this system of orientation, suppose that a crew member wishes to specify the direction of a nearby object or landmark. This is done by indicating the side of the vessel that the object lies on, along with an angle, measured in degrees from the forward part of the ship.

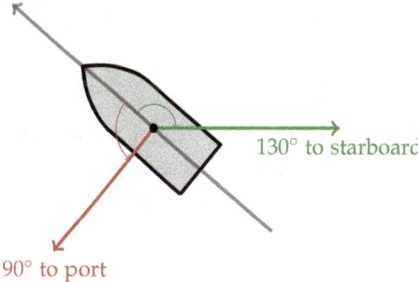

Such a measurement is called the **relative bearing** of the object from the ship, and of course is relative to both the position and orientation of the vessel.

111

LESSON 10

On the other hand, for two ships with different positions and orientations to communicate, it is useful to have a common system. This is where the **compass** becomes useful. A compass always points to the magnetic north pole of the Earth, which becomes a convenient common orientation.

Once the four cardinal directions are found with a compass, bearings can be measured from the northern and southern axes, toward the east or west.

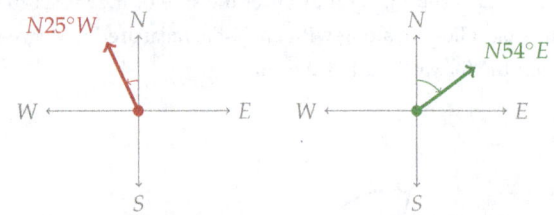

The bearings shown above are read aloud as "25° west of north" and "54° east of north," respectively.

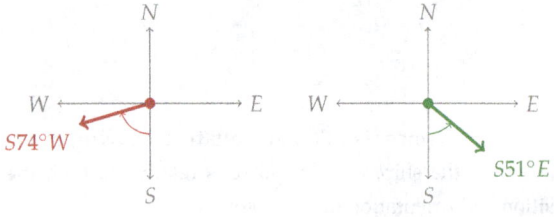

Likewise, the two southward bearings are read aloud as "74° west of south" and "51° east of south," respectively. Bearings indicated with this system are called **magnetic bearings** or **compass bearings**.

So, two different vessels may observe the same fixed object at different compass bearings, depending on their relative **positions**. The orientations of these vessels do **not** play a part in determining these bearings. This is opposed to relative bearings, which depend on a vessel's position and its orientation.

However, if a ship is oriented due north, then the relative bearings and compass bearings may coincide, with the west lying on the port side and the east lying to starboard.

Example: An air traffic controller sees two airplanes on his radar, both flying directly toward his airport. Plane A is at a distance of 25 miles and a bearing of $N61°W$. Plane B is at a distance of 17 miles and a bearing of $N29°E$. What is the relative bearing of Plane A from Plane B?

We are given four numerical values in this problem. We will denote the distances to the two airplanes with Roman letters a and b, and the angles with Greek letters α and β. The unknown relative bearing of Plane A is labeled as θ.

LESSON 10

- $a = 25\,\text{mi}$ (distance to Plane A)
- $b = 17\,\text{mi}$ (distance to Plane B)
- $\alpha = 61°$ (angle of Plane A, from north)
- $\beta = 29°$ (angle of Plane B, from north)
- $\theta = ?$ (relative bearing of Plane A, from Plane B)

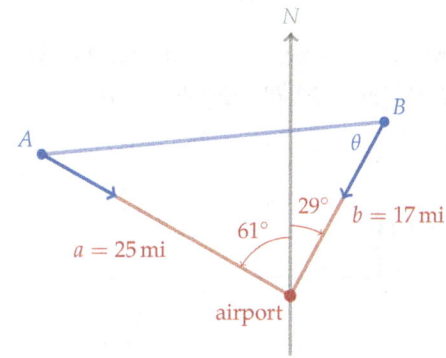

We can see that the sum of the two compass bearings is

$$\alpha + \beta = 61° + 29° = 90°,$$

a right angle, and triangle created by the airport and the two airplanes is a right triangle.

The angle θ, representing the relative bearing we want, has a as its opposite side and b as its adjacent side, so we use the arctangent function

$$\theta = \arctan\left(\frac{25\,\text{mi}}{17\,\text{mi}}\right) \approx 55.78°.$$

When dealing with relative bearings, it is often helpful to rotate our image to adopt the orientation of the vessel measuring the bearing. We've deleted most of the unnecessary labels in the rotated image below; all that we are concerned with at this point is the angle $\theta \approx 55.78°$ and the relative positions of the airplanes. We can see that Plane A is to the starboard side of Plane B.

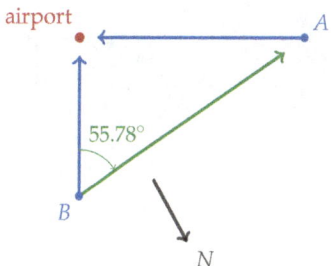

We therefore conclude that Plane A is at a bearing of $55.78°$ to the starboard of Plane B.

113

LESSON 10

Example: A ship leaves port at noon and heads due west at 20 mph. At 2 pm, an island is spotted at 62° to starboard, and they change course to sail directly for it. Find the ship's compass bearing and distance from the port at 3pm.

There are many numerical values given in this problem, but not any distances. We won't be able to make a table of our variables right away, so we'll instead start with the picture.

The ship first leaves the port and travels west for two hours at 20 miles per hour, a total distance of 40 miles. It ends this part of its trip at the point A shown below.

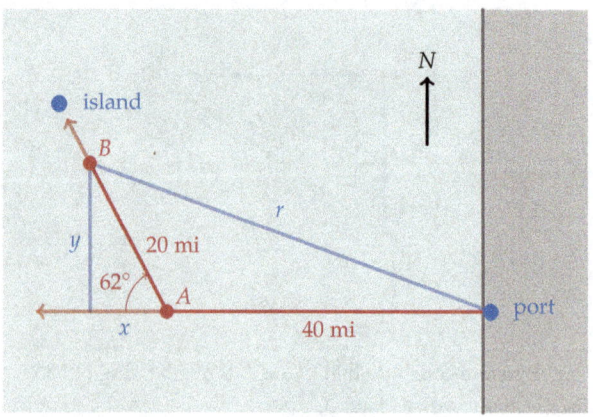

Next the ship changes course from its original path by turning 62° to starboard. In this direction it travels another hour, or 20 miles. In the image, this creates the right triangle with a hypotenuse of 20 miles. We can estimate the lengths of the adjacent and opposite sides:

$$x = 20 \left(\cos 62°\right) \approx 9.39 \,\text{mi},$$

$$y = 20 \left(\sin 62°\right) \approx 17.66 \,\text{mi}.$$

We'll now use these lengths to find both the hypotenuse and angles of the larger blue triangle:

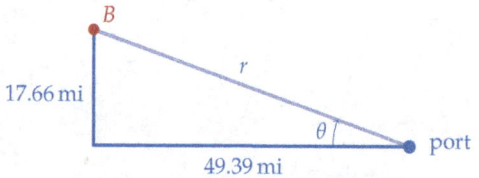

$$r = \sqrt{(49.39)^2 + (17.66)^2} \approx 52.45 \,\text{mi},$$

$$\theta = \arctan\left(\frac{17.66}{49.39}\right) \approx 19.68°.$$

Since θ is here measured from the west, we find that the compass bearing from port to ship is $N70.32°W$, and the distance between the ship and port is about $r = 52.45$ miles.

LESSON 10

EXERCISES

1. Points A and B are on a beach, one mile apart, and P is the top of a distant mountain. All three lie in the same vertical plane. The angle of elevation to P is $10°$ at point A and $25°$ at point B. Draw an accurate scale diagram and use it to estimate the height of the mountain.

2. An airplane flying at 10 thousand feet sees two ships in the water below, both at a bearing of $N30°W$. The angle of depression to one ship is $35°$, and to the other is $20°$. Draw an accurate scale diagram and use it to estimate the distance between the ships.

3. Hillary is standing on top of a 200-foot cliff above a lake. She sees a boat in the water below, and measures the angle of depression to it as $21°$. How far is the boat from the base of the cliff?

4. A navy pilot 1000 feet above the water measures the angle depression to his aircraft carrier to be $63°18'$. What is the diagonal distance between the plane and carrier?

5. A railroad track rises 10 feet for every 400 feet along the track. What is the track's angle of elevation?

6. A tree is broken by the wind. The tree's top touches the ground 13 m away from its base, and makes an angle with the ground of $29°$. How tall was the tree before it broke?

7. The two towers of the Golden Gate Bridge rise 746 feet above the water line. On a small boat nearby, an observer sees an angle of $66°$ between the top of the north tower and its base at the water line. How far is that observer from the base of the north tower?

8. An airplane leaves an airport and flies at 600 miles per hour at a bearing of $S52°W$. After 3 hours, how far south and how far west has the plane traveled from the airport?

9. An airplane leaves an airport and flies due east for 255 miles. It then heads due south for 330 miles. From the plane's current position, what is the relative bearing of the airport, and how far away is it?

10. Two boats leave an island at the same time. Boat A travels due north for 21 miles and Boat B travels due west for 18 miles. What is the compass bearing of Boat A from Boat B? What is the relative bearing of Boat B from Boat A?

LESSON 11

The Law of Sines and the Law of Cosines

Through the last lesson we've gained a bit of experience constructing equations from real-world data. But of course, most triangles that are created by real data will not be **right** triangles, but may take a variety of shapes. If our goal is to solve these triangles, then we'll need to learn some methods that do not depend on the triangles having a right angle.

We've already seen that all triangles can be separated into two right triangles by introducing an **altitude**. By strategically introducing these new line segments and using intermediate lengths and angles, we can typically solve most problems. However, here we'll learn two new laws that apply to **all** triangles, and avoid these intermediate steps.

In terms of their practical usefulness, the laws we learn here could be considered the most important of all the trigonometric laws. We'll illustrate this here with a number of examples, which not only solve a variety of triangles but also determine other physical attributes like area and perimeter.

So far we've made an effort to use Greek letters to denote angles. In this lesson we'll put this convention on hold. Instead, we'll try to emphasize the connection between an angle and its opposing side by using a single Roman letter for both. Each vertex of a triangle has an interior angle associated with it, so the same letter will refer both to the vertex and to the angle it defines.

LESSON 11

THE LAW OF SINES

We start with a general triangle. The triangle in this image has its angles labeled A, B, and C. The sides opposite to each angle are correspondingly labeled a, b, and c.

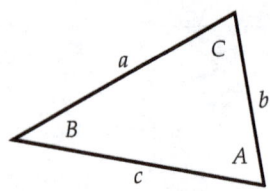

This is not a right triangle, but we separate it into two right triangles by introducing an altitude. This altitude may originate at any of the three angles. We will draw one from the angle C to the side c, intersecting at $90°$.

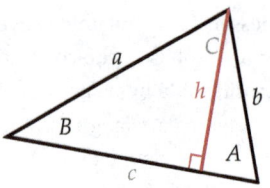

This altitude has a length h, but we will be less concerned with actually finding this length. For us, this is more of an intermediate value that connects the two right triangles in the image. The definition of sine tells us that

$$\sin A = \frac{h}{b} \quad \text{and} \quad \sin B = \frac{h}{a},$$

which we can solve for h. Setting them equal to each other:

$$b \sin A = h = a \sin B.$$

Now we can again ignore the intermediate term h, and divide by the lengths of the sides to obtain the two ratios:

$$\frac{\sin A}{a} = \frac{\sin B}{b}.$$

Since the choice of altitude here was completely arbitrary, we could likewise have drawn altitudes from A to a or from B to b. Each of these would lead to an analogous connection between any of the three ratios.

In fact, it is not even necessary for the triangle in question to have three acute angles. The following is true for any triangle at all, and is referred to as the **Law of Sines**:

$$\frac{\sin A}{a} = \frac{\sin B}{b} = \frac{\sin C}{c}.$$

LESSON 11

The Law of Sines lends itself very well to application problems, as we'll see throughout this lesson. We start with a statement about **isosceles** triangles, which have two sides of equal length.

Example: An isosceles triangle has a base of 22 cm and a vertex angle measuring 36°. Find its perimeter.

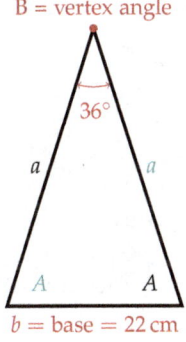

Being an isosceles triangle, the two unknown sides are of equal length a, and their corresponding angles have equal measure A. The equation for perimeter is therefore written

$$p = \text{Perimeter} = (22\,\text{cm}) + 2a.$$

We're now left to find the length of a. This can be done in many ways, but as a first step here we'll actually find the angle A. After all, we know that the three angles add to 180°, so we can solve the following equation:

$$36° + 2A = 180°$$
$$2A = 144°$$
$$A = 72°.$$

Now we use the Law of Sines, connecting the known angles and the lengths of the sides:

$$\frac{\sin B}{b} = \frac{\sin A}{a}$$

$$\frac{\sin(36°)}{22\,\text{cm}} = \frac{\sin(72°)}{a}$$

$$a = \frac{(22)(\sin 72°)}{\sin 36°}\,\text{cm}$$

$$a = \frac{(22)(0.951)}{0.588}\,\text{cm}$$

$$a \approx 35.597\,\text{cm}.$$

Therefore the perimeter is approximately

$$p \approx 22\,\text{cm} + 2(35.597\,\text{cm}) = 93.19\,\text{cm}.$$

LESSON 11

Having just used the Law of Sines to find a perimeter, we'll next find the **area** of a triangle. This solution will not use the Law of Sines per se, but the methods are very similar.

Example: Find the area of a triangle having sides of length 6 ft and 10 ft, if the angle between them is 40°.

We first recall that the formula for the area of a triangle is

$$\text{Area} = \frac{1}{2}(\text{base})(\text{height}).$$

In this formula, the "base" is intended to be one side of the triangle, while the "height" is the length of an altitude that intersects the base. In our image this is the length h.

We draw what we now about our triangle in the image at right. We've labeled the two sides as a and c, respectively, with the angle between them being $B = 40°$. We've also chosen the longest of the three sides as the base. Therefore the altitude creates a smaller right triangle, and:

$$\sin B = \frac{h}{a}.$$

Solving this for h and substituting, we have

$$\text{Area} = \frac{1}{2}(c)(h) = \frac{1}{2}(c)(a \sin B).$$

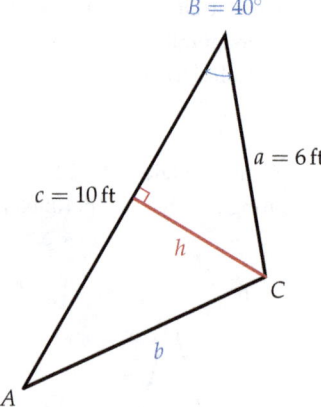

Specifically, our data give us

$$\text{Area} = \frac{1}{2} ac \sin B \approx \frac{1}{2}(6\,\text{ft})(10\,\text{ft})(0.643) = 3.857\,\text{ft}.$$

In the end we've found is the following general formula.

Theorem: Suppose that a triangle has two sides of lengths a and c, with an angle B between them. Then the area of the triangle is given by

$$\text{Area} = \frac{1}{2} ac \sin B.$$

Example: A ship sailing due north sees a lighthouse at 60° to starboard. The ship sails for 10 miles, at which point the lighthouse is at 135° to starboard. How far away is the lighthouse from this second point?

A diagram of this situation is drawn below. We've labeled two points A and B that are 10 miles apart, with the latter being the point where the second bearing of the lighthouse is measured. Note that the ship is sailing north, so the relative bearing of 135° to starboard is also a compass bearing of $S45°E$. The two rays that represent the given bearings intersect at the location of the lighthouse.

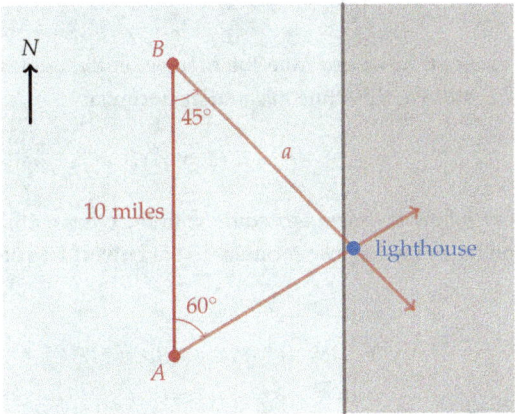

This method of locating a point using measurements from distinct locations is sometimes known as **triangulation**. It is extremely common in surveying and engineering.

Our goal is to find the distance a, and for this we'll use the Law of Sines. We first find the missing third angle:

$$60° + 45° + C = 180°$$
$$C = 180° - 105°$$
$$C = 75°.$$

Now, the Law of Sines says that

$$\frac{\sin C}{c} = \frac{\sin A}{a}$$

$$\frac{\sin(75°)}{10\,\text{mi}} = \frac{\sin(60°)}{a}$$

$$a = \frac{(10)(\sin 60°)}{\sin 75°}\,\text{mi}$$

$$a = (10)\left(\frac{\sqrt{3}}{2}\right)\left(\frac{4}{\sqrt{2}+\sqrt{6}}\right)\,\text{mi}$$

$$a = \frac{20\sqrt{3}}{\sqrt{2}+\sqrt{6}}\,\text{mi}.$$

This exact solution is approximately equal to $a \approx 8.966$ mi.

LESSON 11

THE LAW OF COSINES

To establish our next law, we'll return to a general triangle with angles A, B, and C, and opposing sides labeled a, b, and c. This law will be based more on the Pythagorean Theorem than the Law of Sines, so we'll be focused more on the **lengths** in this triangle.

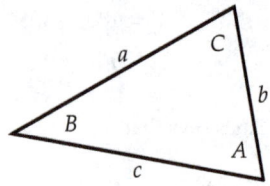

We again separate the triangle into two right triangles with an altitude. This altitude cuts the side that it intersects into two lengths.

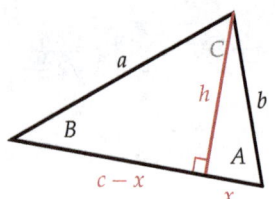

In our image we see an altitude originating at the angle C, which has cut the side c into the lengths x and $c - x$. Now we'll apply the Pythagorean Theorem to these two right triangles, which gives us the equations

$$a^2 = (c - x)^2 + h^2$$
$$b^2 = x^2 + h^2.$$

Substituting the second equation into the first, we simplify:

$$a^2 = (c - x)^2 + (b^2 - x^2)$$
$$= c^2 - 2cx + x^2 + b^2 - x^2$$
$$= b^2 + c^2 - 2cx.$$

For the value x, we see from the triangle on the right that $x = b \cos A$. We substitute this into the formula:

$$a^2 = b^2 + c^2 - 2c\,(b \cos A).$$

We can follow the same procedure with the other altitudes to find the remaining two formulas in the **Law of Cosines**:

$$a^2 = b^2 + c^2 - 2bc \cos A,$$
$$b^2 = a^2 + c^2 - 2ac \cos B,$$
$$c^2 = a^2 + b^2 - 2ab \cos C.$$

Note the similarity of these formulas to the old Pythagorean Theorem we have been using throughout the course. If our triangle were actually a **right** triangle, we might label the right angle as $C = 90°$, and the remaining sides accordingly:

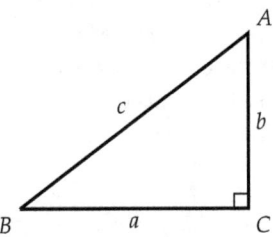

Regardless of the lengths of a or b, because $\cos 90° = 0$, the Law of Cosines now reads

$$c^2 = a^2 + b^2 - 2ab \cos C$$
$$c^2 = a^2 + b^2 - 2ab(0)$$
$$c^2 = a^2 + b^2.$$

So in this sense, we may either think of the Law of Cosines as an extension of the Pythagorean Theorem to general triangles, or think of the Pythagorean Theorem as a special case of the Law of Cosines for right triangles.

Example: In the triangle below, suppose that the sides have lengths $a = 105$ cm, $b = 90$ cm, and $c = 65$ cm. Find the measure of the largest angle.

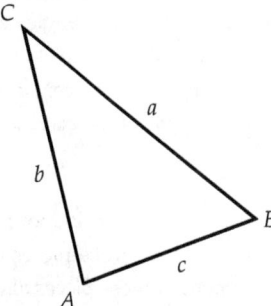

This is a direct application of the formula. The largest angle corresponds to the longest side, so focusing on the angle A:

$$a^2 = b^2 + c^2 - 2bc \cos A$$
$$(105)^2 = (90)^2 + (65)^2 - 2(90)(65) \cos A$$
$$11025 = 4225 + 8100 - 11700 \cos A$$
$$11700 \cos A = 1300$$
$$\cos A = \frac{13}{117}$$
$$A = 83.62°.$$

LESSON 11

SOLVING TRIANGLES

One of the first questions we must answer when approaching a triangle is whether or not we have enough information to successfully **solve** it. For right triangles we have a bit of a head start, since one of their angles is already known. In this case we need only two extra pieces of information: an angle to establish the triangle's similarity class, and a length to fix its size.

For general triangles the situation is more complex. It is still necessary to have at least three pieces of information, but now the arrangement of those pieces also plays a part.

In Lesson 3 we learned that knowing all three **angles** of any triangle is enough to determine its shape, or similarity class. We sometimes abbreviate this situation by writing **AAA**, meaning that we know all three angles of a given triangle. But without knowing the lengths of any of its sides, it would still be impossible to solve such a triangle.

On the other hand, if we knew the lengths of all three **sides**, as we did in the last example, then we can use the Law of Cosines to determine all three angles as well. We refer to this situation as **SSS**. In every such case the triangle is able to be completely and uniquely solved for all of its attributes.

Next we turn to combinations of sides and angles, when their arrangement on the triangle becomes non-trivial.

Example: Suppose that a triangle has two sides of length $a = 3$ ft and $b = 2$ ft, and an angle between them of $B = 80°$.

Since the angle lies in between the two sides, we refer to this situation as "side-angle-side," or **SAS**. We can see from the image that the free endpoints of the line segments a and c can be naturally connected by a third line segment b.

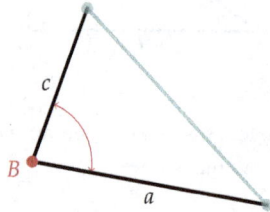

In fact, the length of the third side is uniquely determined by the Law of Cosines:

$$b^2 = (3)^2 + (2)^2 - 2(3)(2)\cos 80° \approx 10.916$$
$$b \approx 3.304 \text{ ft}.$$

Now that all three sides lengths are known, we are in the situation **SSS**, and the triangle is able to be completely solved.

LESSON 11

Example: Let one side of a triangle have length $b = 5\,\text{yd}$, and the angles at the endpoint measures $A = 35°$ and $C = 60°$, respectively.

In this example the side length lies between the two angles, so we refer to this situation as "angle-side-angle," or **ASA**.

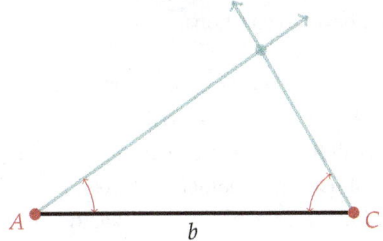

We draw rays from each endpoint making the prescribed angles with the line segment b. We see that the two rays intersect at a single point. Since we know the angles A and C, we also know the angle at this vertex, $B = 85°$. We can now use the Law of Sines to determine the remaining side lengths. This is exactly the same process that we saw in our earlier example, with the ship that was sailing due north.

In general, as long as the sum of the two angles is less than $180°$, the two rays will intersect and we will always be able to solve the triangle uniquely.

Example: Suppose that a triangle has a side of length $b = 4\,\text{cm}$ and an opposing angle of $B = 50°$. Let one of the remaining angles have the measure $A = 20°$.

Two angles are given here, but not the side length between them. Instead the side opposite to B is given, so this arrangement is known as "side-angle-angle," or **SAA**.

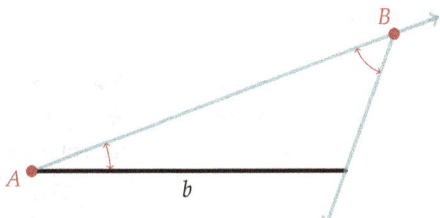

This is a new arrangement, but it is actually not significantly different from the previous case. In both of these examples we are given two angles, and if their sum is less than $180°$, it is simple to find the third. At that point we may use the Law of Sines to find the two missing side lengths. For example, the length a is given by

$$\frac{\sin 50°}{4\,\text{cm}} = \frac{\sin 20°}{a}$$

$$a = \frac{4\sin 20°}{\sin 50°}\,\text{cm} \approx 1.786\,\text{cm}.$$

LESSON 11

Our only remaining situation involves having one angle and two sides, in which the known angle does not lie between the two sides. This situation is referred to (colloquially) as **ASS**. Here we will not give an explicit example, as there are many possible outcomes.

Suppose that we are given adjacent side lengths a and b, and an angle opposite to one of the sides, A. For now, also assume that $A \geq 90$.

In the image below we can see an obtuse angle A, along with two sides having the property that $a \leq b$ It is clear no such triangle can exist, as the line segment a will never intersect the blue ray.

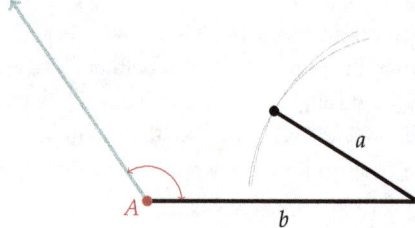

We can also see this algebraically with the Law of Cosines:

$$a^2 = b^2 + c^2 - 2bc \cos A.$$

Moving the b^2 term to the other side, we have:

$$a^2 - b^2 = c^2 - 2bc \cos A.$$

Since $a \leq b$, we see that the term on the left is **negative**. However, because $A \geq 90°$, we know that $\cos A \leq 0$. So the right-hand side is **positive**, a contradiction. Therefore this equation can have no solutions.

However, if $a > b$, then both sides of the above equation are positive. In this case the equation will have exactly one solution c; recall that the values a, b, and A are all given. We can see this illustrated geometrically below, as the longer side a must intersect the blue ray in exactly one place.

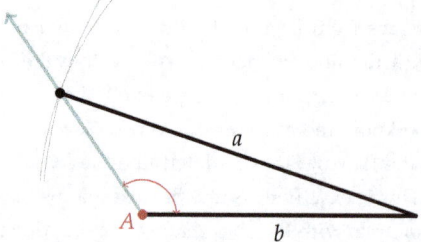

Again, actually solving the triangle uses the Law of Cosines to find the length of the third side c, which then puts us in the situation **SSS**.

LESSON 11

Next, suppose that $A < 90$. Below we see a triangle with an acute angle A, a side b, and a longer side a. It is clear that if $a > b$, then it must intersect the blue ray in exactly one place, creating an obtuse angle C.

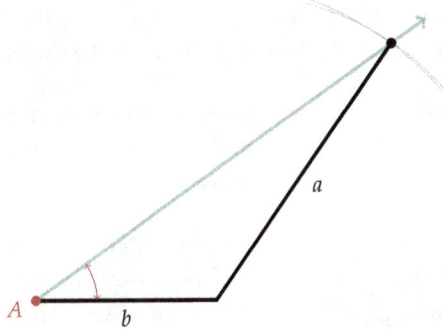

However, if a is too short, then it will not intersect the blue ray at all. Thus there is no triangle satisfying the conditions.

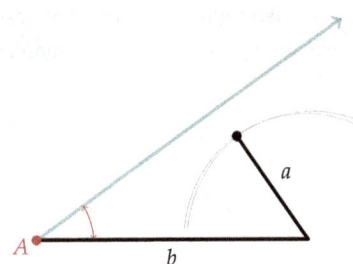

Now, suppose that a is a longer segment, but still having $a < b$. Then it will intersect the blue ray in two places, creating two possible triangles.

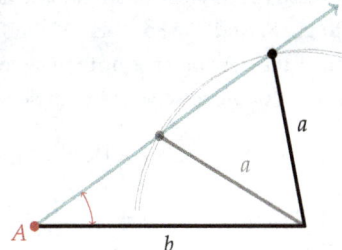

Finally, if a intersects the blue ray **tangentially**, then this intersection will occur at a right angle. This creates a unique **right** triangle, with b as its hypotenuse. Therefore we know that $a = b \sin A$. This is the boundary that separates the conditions for acute angles A:

- If $b < a$, then a unique triangle exists. The missing side is found with the Law of Cosines.

- If $b \sin A < a < b$, then there are two such triangles. The missing sides are found with the Law of Cosines.

- If $a < b \sin A$, then no such triangle exists.

- If $a = b \sin A$, then a unique right triangle exists. The missing side c is found with the Pythagorean Theorem.

LESSON 11

EXERCISES

Suppose that a triangle has sides of length a, b and c, and angles of length A, B, and C. For each of the following, determine whether the Law of Sines or the Law of Cosines would be more appropriate to solve the triangle.

1. $a = 10$, $A = 40°$, $c = 8$
2. $a = 14$, $b = 15$, $c = 16$
3. $c = 21$, $a = 14$, $B = 60°$
4. $b = 17$, $B = 42°58'$, $a = 11$
5. $c = 14.1$, $A = 29°$, $b = 7.6$
6. $A = 28°50'$, $b = 5$, $c = 4.9$

7. A 40-foot antenna stands on top of a building. From a point on the ground, the angles of elevation to the top and bottom of the antenna are 56° and 42°, respectively. How tall is the building?

8. Two planes leave an airport at the same time, each flying at 110 miles per hour. One flies at a bearing of $N60°E$. The other flies at a bearing of $S40°E$. How far apart are the planes after 3 hours?

9. The sides of a triangle have lengths 50 ft, 70 ft, and 85 ft. Find the measure of the smallest angle.

10. Frank is visiting the space shuttle, and from where he stands he measures the angle of elevation to its top as 27.2°. Then he walks 17.5 meters further away from the shuttle, and measures 23.9°. What is the height of the shuttle?

11. A triangular lot faces two streets that meet at an angle of 85°. The sides of the lot facing the streets are 160 feet in length each. Find both the area and perimeter of the lot.

12. The towers of the Golden Gate Bridge are 4200 feet apart. An observer at the north tower sees a small boat that forms an angle of 37° with the south tower. An observer at the south tower sees the same boat form an angle of 43° with the north tower. How far is the boat from each tower?

In Exercises 13–18, determine the number of possible triangles with the following properties. If a solution exists, solve the triangle.

13. $a = 6$, $b = 10$, $A = 36°52'$
14. $a = 7$, $b = 6$, $A = 30°$
15. $b = 40$, $a = 32$, $A = 125°20'$
16. $a = 26$, $b = 29$, $A = 58°$
17. $A = 25°$, $a = 125$, $b = 150$
18. $A = 76°$, $a = 5$, $b = 20$

LESSON 12

Coordinate Systems and Graphs

We've already spent a good amount of time investigating the trigonometric functions and their role in the measurement of lengths and angles. At the same time, we understand that in actual applications these measurements will usually take place in three spatial dimensions. Thus we'll need to transfer what we've learned to three dimensions. The first step of this is to establish a coordinate system.

Over the next few lessons we will develop several standard ways of doing this. Our first coordinate system was established in Lesson 5. That system included two spatial dimensions, and each point P was represented by a unique ordered pair $P = (x, y)$.

Because of this, we often call this space the x, y-plane, or further abbreviate it with the shorthand symbol $\mathbb{R}^2$. We can create a similar system for three dimensions by assigning to each point in space an **ordered triple**, $P = (x, y, z)$. The resulting set of points is then known as **x, y, z-space**, and abbreviated $\mathbb{R}^3$.

These coordinates will give us a framework in which to measure and compare geometric objects like curved surfaces, which do not necessarily lie in a single flat plane. We proceed much like we did in Lesson 5. Many of the definitions and formulas will be completely analogous to their two-dimensional counterparts.

LESSON 12

Lesson 5 taught us that every coordinate system consists of the following choices:

1. A choice of **origin**.
2. A standard unit of **distance**.
3. A choice of **orientation** for each dimension.

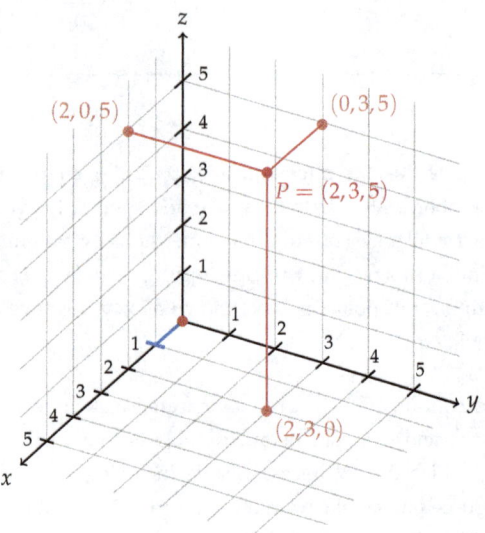

In the image at left, we've begun with an origin and a unit of distance. We've also used the standard orientation for the three coordinate axes, with only the **positive** parts of each axis shown. All formulas and geometric objects that we discuss below will be based on this orientation.

Note that this is a two-dimensional representation of a three-dimensional space; each pair of axes is to be interpreted as meeting at a right angle. Note also that a point $P = (2,3,5)$ is indicated, along with three other, unnamed points. We'll use these to illustrate a few definitions.

Definition: The three planes formed by each pair of axes are called the **coordinate planes**. Specifically, the plane formed by the x- and y-axes is called the x,y-plane. Likewise, the plane formed by the y- and z-axes is called the y,z-plane, and the plane formed by the x- and z-axes is called the x,z-plane.

Definition: Suppose that $P = (x,y,z)$ is a point in a three-dimensional Cartesian coordinate system. The point $P_{xy} = (x,y,0)$ is called the **projection of P onto the x, y-plane**. Likewise, the point $P_{yz} = (0,y,z)$ is called the projection of P onto the y,z-plane, and $P_{xz} = (x,0,z)$ is called the projection of P onto the x,z-plane.

LESSON 12

We can see that for a point P in $\mathbb{R}^3$ defined by an ordered triple $P = (x, y, z)$, the coordinate x indicates the horizontal distance of the point from the y, z-plane. Likewise, y indicates the horizontal distance of the point from the x, z-plane, and z indicates the vertical distance of the point from the x, y-plane.

When we established the coordinate system in two dimensions, the two coordinate axes separated the plane into four quadrants. Here, the three coordinate planes separate the space into eight **octants**. That is, there are eight different combinations of signs that a point in space could have.

For consistency with the quadrants of $\mathbb{R}^2$, we name the four octants with positive z-coordinates in the same manner. That is, the point $P = (2, -8, 5)$ lies in Octant IV of $\mathbb{R}^3$, because its projection $P_{xy} = (2, -8, 0)$ lies in Quadrant IV of the x, y-plane. Similarly, $Q = (-1, -1, 9)$ lies in Octant III.

For negative z-coordinates, we use the same orientation, starting from the Octant IV which lies directly below Octant I. The point $Q = (-6, 1, -7)$ thus lies in Octant VI, because its projection $Q_{xy} = (-6, 1, 0)$ lies in Quadrant II of the x, y-plane. Octants II and VI lie adjacent to each other, across the x, y-plane.

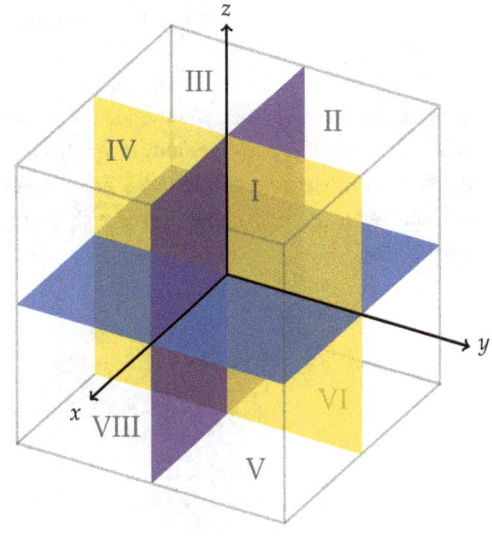

Octant I: $(+,+,+)$ Octant V: $(+,+,-)$

Octant II: $(-,+,+)$ Octant VI: $(-,+,-)$

Octant III: $(-,-,+)$ Octant VII: $(-,-,-)$

Octant IV: $(+,-,+)$ Octant VIII: $(+,-,-)$

Lesson 12

With the addition of this new coordinate system we can now discuss geometry in both two and three dimensions. We'll start with some general ideas.

The most basic notion of geometry is the idea of distance. We already have a lot of experience finding lengths of line segments. Now let us suppose that P and Q are two points in a coordinate system. An example of two such points in $\mathbb{R}^2$ is shown below.

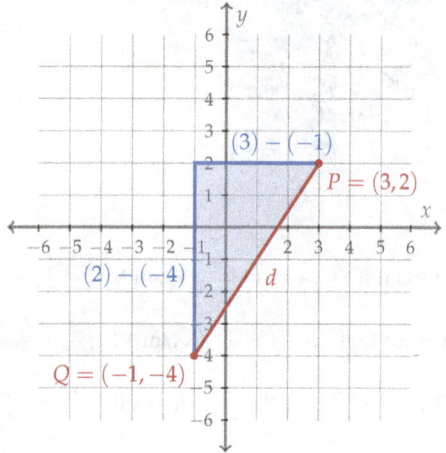

To find the distance between them, we first draw the line segment that joins them, which is labeled as having a length d. We can also create vertical and horizontal line segments based on the coordinates of the points, forming a right triangle. This reduces our problem to something very familiar.

We next determine the lengths of the vertical and horizontal segments through the coordinates of the points. Shown in the image, we subtract the x-coordinates of the points to find the length of the horizontal segment, and the same for the vertical segment. Applying the Pythagorean Theorem:

$$d = \sqrt{(3-(-1))^2 + (2-(-4))^2} = \sqrt{16+36} = \sqrt{52}.$$

We can use this to define a new function for distance.

Definition: Given two points $P = (x_1, y_1)$ and $Q = (x_2, y_2)$ in $\mathbb{R}^2$, the **distance** between them is given by

$$\text{dis}(P, Q) = \sqrt{(x_1 - x_2)^2 + (y_1 - y_2)^2}.$$

Likewise, given two points $P = (x_1, y_1, z_1)$ and $Q = (x_2, y_2, z_2)$ in $\mathbb{R}^3$, the **distance** between them is given by

$$\text{dis}(P, Q) = \sqrt{(x_1 - x_2)^2 + (y_1 - y_2)^2 + (z_1 - z_2)^2}.$$

Obtaining the function just given for $\mathbb{R}^3$ uses a very similar method, and we will see this in practice in Lesson 15. For now, let's continue working in $\mathbb{R}^2$ to identify another geometric object, the midpoint between two points.

Here the main idea is to work with each coordinate individually. Two new points P and Q are shown in the image at right. These have x-coordinates of -2 and 5, respectively. Finding the midpoint between these two numbers on the x-axis leads us to take their **average**. We do the same for y:

$$x_{av} = \frac{(-2)+(5)}{2} = \frac{3}{2} \quad \text{and} \quad y_{av} = \frac{(4)+(-1)}{2} = \frac{3}{2}.$$

This gives the coordinates of the midpoint $M = \left(\frac{3}{2}, \frac{3}{2}\right)$.

Definition: Given two points $P = (x_1, y_1)$ and $Q = (x_2, y_2)$ in $\mathbb{R}^2$, their **midpoint** is given by

$$\text{mid}(P, Q) = \left(\frac{x_1 + x_2}{2}, \frac{y_1 + y_2}{2}\right).$$

Likewise, given two points $P = (x_1, y_1, z_1)$ and $Q = (x_2, y_2, z_2)$ in $\mathbb{R}^3$, their **midpoint** is given by

$$\text{mid}(P, Q) = \left(\frac{x_1 + x_2}{2}, \frac{y_1 + y_2}{2}, \frac{z_1 + z_2}{2}\right).$$

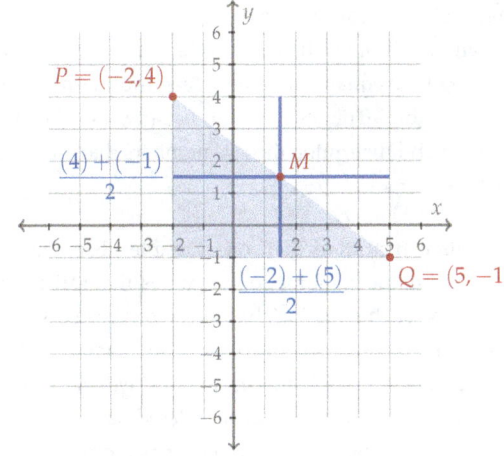

Again, we've used an example in $\mathbb{R}^2$ for simplicity, but the formula is no different for points in three dimensions. To find midpoints, we average each of the coordinates individually. These averages form the coordinates of the midpoint. This distance function outputs a **positive real number**. This midpoint function outputs a **point in space**.

Coordinate systems provide a global frame of reference in which we can specify position and compare objects, but also create a link between two dissimilar mathematical ideas.

Lesson 12

By itself, an **equation** is just an algebraic expression involving an equals sign. It may have any number of variables. Most of the equations we have seen thus far have had only one variable. This means that, usually, we can solve the equation and determine the real numbers that make the equation true.

If an equation has more than one variable, then there is typically no unique combination of values that satisfy the equation. One such equation that we have repeatedly seen is $x^2 + y^2 = 1$. Of course, there are infinitely many combinations of x and y that satisfy this equation. If we choose such a combination, we can write it as an ordered pair (x, y), and once written this way, we can consider it as a **point**, which is a geometric object, and place it at its appropriate position in the coordinate system.

Definition: The set of points whose coordinates satisfy a given equation is called the **graph** of the equation.

Apart from the Unit Circle itself, we can use these ideas to find the equation of a circle of any radius. A circle is usually defined, geometrically, as the set of points that are a fixed distance from a center. Let's choose a point to use as an example, like $P = (-1, 2)$, and a radius, $r = 4$.

The formula for distance in $\mathbb{R}^2$ was

$$\text{dis}(P, Q) = \sqrt{(x_1 - x_2)^2 + (y_1 - y_2)^2}.$$

Here our distance is fixed, as is one of the two points, so we can fill in these values. It does not matter which x, y pair we use to substitute. Since we have only one point that remains variable, we remove the subscripts on those coordinates.

$$\sqrt{((-1) - x)^2 + ((2) - y)^2} = 4$$
$$(-1 - x)^2 + (2 - y)^2 = 16$$
$$(1 + 2x + x^2) + (4 - 4y + y^2) = 16$$
$$x^2 + 2x + y^2 - 4y = 11.$$

The following gives both a geometric and algebraic definition of the same object.

Definition: Let $P = (h, k)$ be any point in $\mathbb{R}^2$, and let $r > 0$ be any real number. The **circle** with center P and radius r is defined as the set of points that are the distance r away from P. According to the distance formula, the coordinates of a point on the circle will satisfy the equation

$$(x - h)^2 + (y - k)^2 = r^2.$$

LESSON 12

We can also form an analogous definition for an object in three dimensions, again coming from the distance formulas.

Definition: Let $P = (x_1, y_1, z_1)$ be any point in $\mathbb{R}^3$, and let $r > 0$ be any real number. The **sphere** with center P and radius r is defined as the set of points that are the distance r away from P. The coordinates of a point on the sphere will satisfy the equation

$$(x - x_1)^2 + (y - y_1)^2 + (z - z_1)^2 = r^2.$$

Example: Let $P = (1, 4, 3)$, and $r = 2$. Then all points on the sphere have coefficients that satisfy the equation:

$$(x - 1)^2 + (y - 4)^2 + (z - 3)^2 = 2^2$$
$$(x^2 - 2x + 1) + (y^2 - 8y + 16) + (z^2 - 6z + 9) = 4$$
$$x^2 - 2x + y^2 - 8y + z^2 - 6z = -22.$$

The first person to connect geometry and algebra with these coordinate systems was René Descartes. The coordinate systems themselves were first described in his work **La Géométrie**, published in 1637. For this reason, the coordinate systems we've seen so far are known as **Cartesian** coordinate systems. Descartes later used these to found the branch of mathematics known as **analytic geometry**.

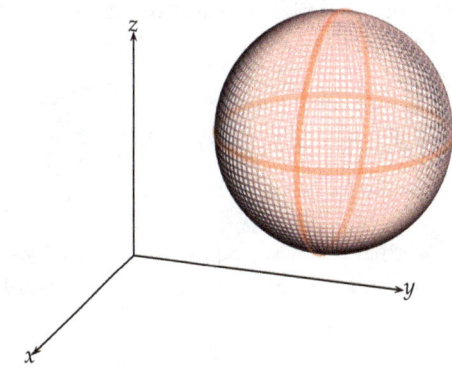

In the next lesson we'll learn alternate ways to identify points and graphs in $\mathbb{R}^2$ and $\mathbb{R}^3$. When discussed in opposition to other coordinate systems, the Cartesian coordinates are sometimes known as **rectangular** coordinates.

Circles and spheres are good examples of graphs, and we will be exploring their equations in greater detail soon. But they are certainly not the only examples. For contrast, we'll next look at some graphs that are "flat" in their respective spaces. A definition will help make this precise.

Definition: The **degree** of a variable in a given equation is the highest exponent in which that variable appears.

135

LESSON 12

In these previous examples, the equations of both the circle and sphere contained terms that, at maximum, had degree two:

$$x^2 + 2x + y^2 - 4y = 11,$$

$$x^2 - 2x + y^2 - 8y + z^2 - 6z = -22.$$

When we say that our next graphs will be "flat," we can make this precise by saying that their equations will have only variables of degree one.

Examples: The equations $x = 2$ and $y = 3$ are both equations whose variables have degree one. In this case we usually say that the equations themselves have degree one.

We'll first draw the graphs of these equations in $\mathbb{R}^2$. In the first case, the graph will be the set of points (x, y) satisfying $x = 2$. We are not given any information about the y-coordinate, or its relation to x. We could write this set in set-builder notation as

$$M = \{(2, y) \in \mathbb{R}^2 \mid y \in \mathbb{R}\}.$$

Some points that lie in this set are $(2,0)$ and $(2,3)$, though there are infinitely many others. The set of all such points forms a **vertical line**, shown in red at right.

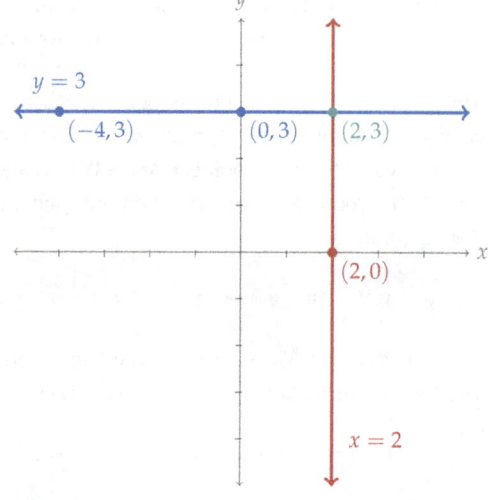

The points satisfying the second equation, $y = 3$, can likewise be written as a set:

$$N = \{(x, 3) \in \mathbb{R}^2 \mid x \in \mathbb{R}\}.$$

In this case the points all have the same y-coordinate. A few such examples are $(0,3)$ and $(-4,3)$. However, the collection of all points in T forms a **horizontal line**, parallel to the x-axis.

LESSON 12

Of course, there are also examples of lines that are not perfectly vertical or horizontal. We've already seen many rays that originate at the origin and extend in a certain direction. All of these may be further extended into a straight line.

Any straight line that passes through the origin makes a fixed angle with the positive x-axis, and therefore the **tangent** of that angle is also fixed. We call this value, $m = \tan\theta$, the **slope** of the line.

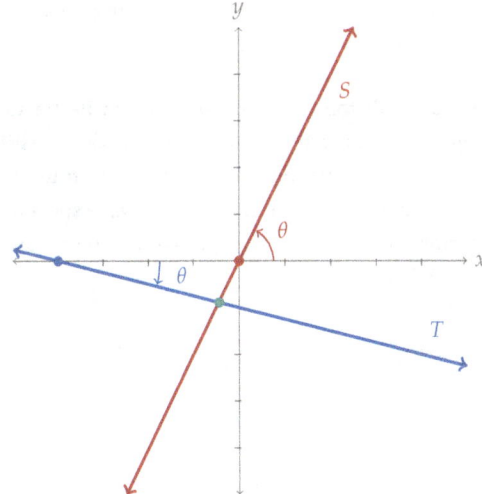

Because each line extends to infinity in both directions, we can always consider these angles θ to be on the interval $(-90°, 90°)$. This is advantageous as the tangent function is one-to-one on this fundamental domain.

A more general line that does not pass through the origin will still have an intersection with the x-axis at a specific angle, and so will also have a slope. In this situation we can find the slope by selecting any two points on the line.

Definition: Given two points (x_1, y_1) and (x_2, y_2) in $\mathbb{R}^2$, the **slope** of the line between them is

$$m = \frac{y_2 - y_1}{x_2 - x_1} = \frac{\text{"rise"}}{\text{"run"}}.$$

If both points have the same x-coordinate, so that the line passing through them is vertical, then we say that the slope is **undefined**.

Definition: Two lines are called **parallel** if their slopes are equal. Two lines are called **perpendicular** if the product of their slopes is equal to -1.

It is clear from the image at left that the lines shown are neither parallel nor perpendicular.

137

LESSON 12

Example: Find the slope of the line $5x - 2y = 30$. Estimate the angle θ that the line makes with the positive x-axis.

For this we'll want to recall both of the definitions of slope that we have seen so far:

$$\tan\theta = m = \frac{y_2 - y_1}{x_2 - x_1}.$$

To use this formula on the right we will need to locate two points that lie on the line. Choosing a specific value for either x or y will leave us with an equation in only one variable, which we will be able to solve. We can choose any values we like for this purpose. To make the computations easier, we will choose to set $x = 0$:

$$5(0) - 2y = 30$$
$$-2y = 30$$
$$y = -15.$$

We also set $y = 0$ to get a second point:

$$5x - 2(0) = 30$$
$$5x = 30$$
$$x = 6.$$

This tells us that the combinations $(x_1, y_1) = (0, -15)$ and $(x_2, y_2) = (6, 0)$ both satisfy the equation, so that these points lie on the line. From the second definition:

$$m = \frac{(0) - (-15)}{(6) - (0)} = \frac{5}{2}.$$

To find the angle θ, we use the first definition:

$$\theta = \arctan(2.5) \approx 68.2°.$$

Example: Find the equation of the line having slope $m = 3$, and passing through $P = (-4, 2)$.

We can start with the definition of slope in its fractional form. Since we know the value of m, we can substitute it directly into the equation. We can also substitute the x and y-coordinates of P into one of the available spots in the denominator and numerator, respectively. The remaining variables are left blank.

$$(3) = \frac{y - (2)}{x - (-4)}$$
$$3(x + 4) = y - 2$$
$$3x + 12 = y - 2$$
$$3x - y = -14.$$

LESSON 12

Now that we've looked at some lines and their slopes we can turn to sets of lines. In a plane, any two lines that are not parallel must intersect at a point. We can identify such a point geometrically without much problem, but we would like to be able to find it algebraically as well.

Definition: The **intersection** of two graphs is the set of points that they have in common. That is, if S and T are two graphs, then their intersection is $S \cap T$.

In practice, we find these points of intersection by solving a system of equations. There are several methods of doing this, and we'll illustrate one such method with an example.

Example: The following two lines are not parallel. Find their point of intersection.

$$3x - 5y = 2$$
$$6x + y = -1$$

We begin by solving one of the equations for one of its two variables. Since the y-term of the second equation has a coefficient of one, it will make a good candidate:

$$y = -6x - 1.$$

Next, we substitute the expression on the right into the other equation. This will result in an equation that has only one variable, which we can solve uniquely.

$$3x - 5y = 2$$
$$3x - 5(-6x - 1) = 2$$
$$3x + 30x + 5 = 2$$
$$33x = -3$$
$$x = -\frac{1}{11}.$$

Finally, we'll substitute this value into either of the two equations to find the y-coordinate. Note that this again gives us a unique solution. We should expect this; the two lines intersect in exactly one point.

$$6\left(-\frac{1}{11}\right) + y = -1$$
$$y = -1 + \frac{6}{11}$$
$$y = -\frac{5}{11}.$$

Therefore the point of intersection is $(x, y) = \left(-\frac{1}{11}, -\frac{5}{11}\right)$.

139

Lesson 12

Finally, we'll look at some three-dimensional graphs that are "flat," meaning that their equations have degree one.

Example: Consider again the equations $x = 2$, $y = 3$, but this time as equations relating three variables x, y, and z. This time, the first equation defines a set of points in $\mathbb{R}^3$ having x-coordinate equal to 2, or:

$$P = \{(2, y, z) \in \mathbb{R}^3 \mid y, z \in \mathbb{R}\}.$$

This set of points forms a flat vertical plane, parallel to the y, z-plane, which is shown in red in the image at right.

One way to describe this plane is to start with the **line** M in the x, y-plane, defined by the equation $x = 2$, as we did previously. Since the variable z does not appear in this equation, each point $(2, y, 0)$ on M actually defines a vertical line passing through it. For example, the point $(2, 3, 0)$ lies on the line M, but also generates the vertical line

$$L = \{(2, 3, z) \in \mathbb{R}^3 \mid z \in \mathbb{R}\}.$$

In this context we could say that the line M **generates** the plane P, or that M is the **projection** of P into the x, y-plane. In any case, M is the intersection of the plane P and the plane $z = 0$.

Similarly, the second equation $y = 3$ defines the set

$$Q = \{(x, 3, z) \in \mathbb{R}^3 \mid x, z \in \mathbb{R}\}.$$

These points again form a vertical plane, shown in blue.

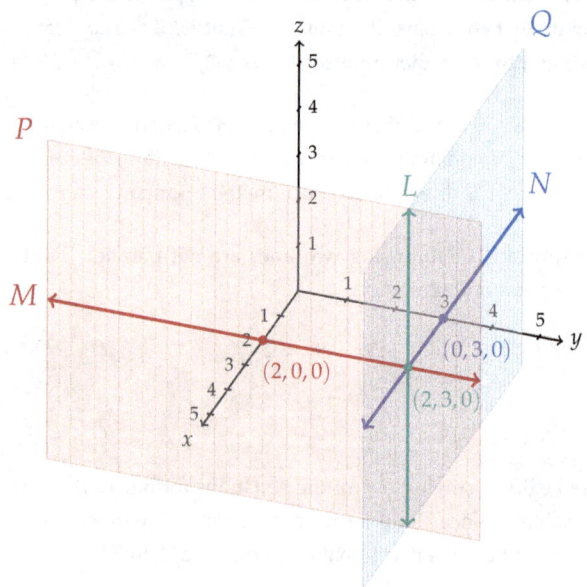

This plane is generated by the line

$$N = \{(x, 3, 0) \in \mathbb{R}^3 \mid x \in \mathbb{R}\}.$$

Since $(2, 3, 0)$ is the point of intersection of the lines M and N, the vertical line L is common to both P and Q, and thus forms the intersection of those two planes. That is,

$$L = P \cap Q.$$

Just as two non-parallel lines will always intersect at a point, two non-parallel planes will always intersect in a line.

Example: Sketch a graph of the plane $z = x$ in $\mathbb{R}^3$.

The plane is defined to be the set of points satisfying this equation. When we write the set, we will make this specification is the description of the points:

$$V = \{(x, y, x) \in \mathbb{R}^3 \mid x, y \in \mathbb{R}\}.$$

For example, the point $(2, 4, 2)$ is in S, but the point $(1, 2, 3)$ is not, because the x- and z-coordinates are not equal. The full collection of points is shown in red in the image at right. The entire y-axis lies in the plane, as the points $(0, y, 0)$ all satisfy our equation.

We could also consider the line U in the x, z-coordinate plane that is defined by the same equation, $z = x$. This is a diagonal line of slope one, and each point on it generates a horizontal line in $\mathbb{R}^3$ that is parallel to the y-axis.

Again, in this context we could say that the line U **generates** the plane V, or that U is the **projection** of V in the x, z-plane.

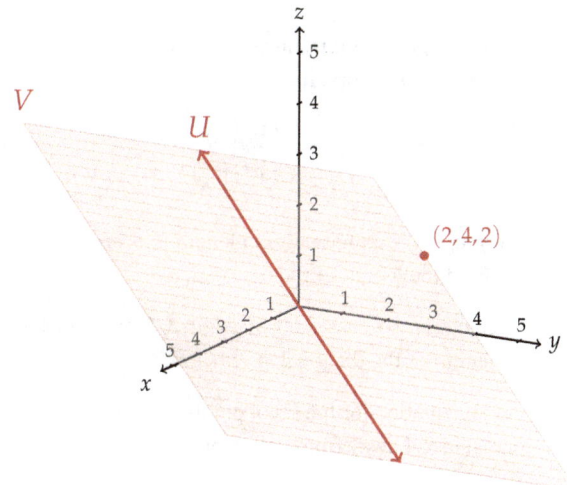

LESSON 12

EXERCISES

In Exercises 1–9, determine the octant in which each point lies, and plot each point in a single coordinate system.

1. $(1, 2, 1)$
2. $(-3, -3, -1)$
3. $(-1, 3, 4)$
4. $(-2, 3, -5)$
5. $(2, 1, -1)$
6. $(1, -2, 1)$
7. $(3, -1, -2)$
8. $(5, -2, 7)$
9. $(-4, -7, 2)$

For each pair of given points, find both the distance between the points and their midpoint.

10. $(12, 6, 1)$, $(8, 2, -3)$
11. $(-2, 5, 7)$, $(1, 2, 3)$
12. $(2, 1, -1)$, $(-2, -1, 1)$
13. $(15, 4, -7)$, $(-9, 0, 3)$

14. A rectangular box has sides of length 6, 8, and 10 centimeters. Find the length of the diagonal, d.

15. A rectangular box has sides of length 3, 5, and 7 inches. Find the length of the diagonal, d.

16. Find an equation for the set of points (x, y) in the plane that are distance 4 away from the point $(1, -1)$.

17. Find an equation for the set of points (x, y, z) in space that are distance 3 away from the point $(1, -2, 3)$.

18. In the plane, find the slope of the line passing through the points $(-4, 6)$ and $(-1, 0)$.

19. In the plane, find the slope of the line passing through the points $(-3, -1)$ and $(3, 3)$.

20. Write an equation of the line passing through the point $(0, -7)$ that has slope $m = -2/5$.

21. Write an equation of the line passing through the point $(-3, -2)$ that has slope $m = 3/4$.

Is each pair of lines parallel, perpendicular, or neither?

22. $4x - y = 3$, $12 - 3y = 3$
23. $3x - 10y = 7$, $5x + 4y = 0$
24. $3x + y = 4$, $2x = 6y + 8$
25. $4y + x = 17$, $2x + 4y = 1$

26. Find the point of intersection of the lines $3y - 2x = 6$ and $x - 2y = -5$.

27. Find the point of intersection of the lines $2y - 3x = 19$ and $x + 2y = 7$.

LESSON 13

Polar Coordinates

In both Lesson 5 and Lesson 12 we saw that the standard coordinate system for a two or three-dimensional space are the Cartesian coordinates. While these coordinates work very well for most purposes, there are other methods one can use to identify position, and we will explore one of them now: **polar coordinates**.

This set of coordinates is particular to the two-dimensional plane $\mathbb{R}^2$. Previously, we specified two perpendicular axes, which created an origin at their intersection. After deciding on a unit of measurement, we then identified a point by its distance to each axis.

Now, we will instead fix both an origin and a specific direction. A point P in space will then be identified by a distance and an angle in relation to this fixed point and direction.

Definition: Let O and D be two fixed points in the plane, so that the line segment from O to D fixes both a unit of distance and a direction in the plane. Then let P be any other point in the plane. The length r of the line segment from O to P is called the **radial coordinate** or **radius** of P. The angle θ created by the two line segments is called the **angular coordinate** or **polar angle** of P. We then denote P as $P = (r, \theta)$.

LESSON 13

Here the point O is the origin of the new coordinate system. The point D provides us not only with a unit of distance, but also with a line segment against which we can measure angles. The choice of **orientation** now consists of the direction in which we measure these angles. To be consistent with our previous work, we will measure all angles in the **counterclockwise** direction.

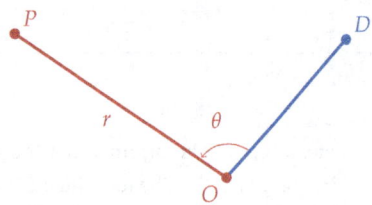

We'd like to connect this to the Cartesian coordinate system we've already established, so suppose that we already have a choice of origin $O = (0,0)$. This will remain the origin of our polar coordinate system.

The old Cartesian system also had a unit of distance that allowed us to mark both axes at regular intervals. To maintain this distance, we place the point D at the position $(1,0)$, on the **positive x-axis**.

As always, we'll measure angles **counterclockwise** from the positive x-axis, in the direction of the positive y-axis. Therefore the angular coordinate of the point P in the image below will be **obtuse**, rather than the acute angle measured in the clockwise direction.

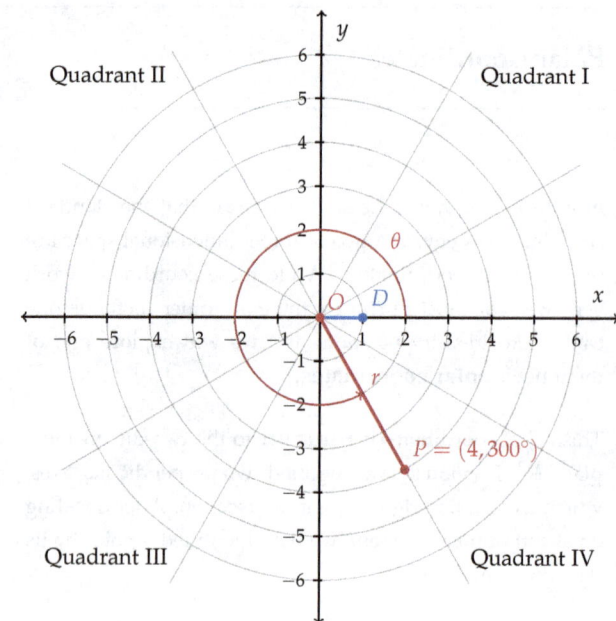

Example: In the previous image, we saw a point P that was labeled in polar coordinates as $P = (4, 300°)$. We will find the Cartesian coordinates of P.

First, we look at the reference triangle created in Quadrant IV by the point P. This triangle has a reference angle of $60°$ with the positive x-axis. Also, the x-coordinate of P is the length of the adjacent side of this triangle.

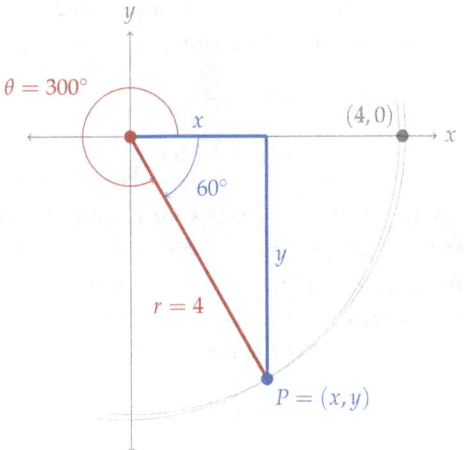

We find the x-coordinate using the hypotenuse $r = 4$, the reference angle, and the definition of cosine:

$$\frac{x}{r} = \cos \theta$$

$$x = r \cos \theta$$

$$x = (4) \cos(60°)$$

$$x = (4)\left(\frac{1}{2}\right)$$

$$x = 2.$$

We use the sine function to find the y-coordinate of P. The reference angle will give us a positive distance for y, which we later make negative to place P in the correct quadrant.

$$\frac{y}{r} = \sin \theta$$

$$y = r \sin \theta$$

$$y = (4) \sin(60°)$$

$$y = (4)\left(\frac{\sqrt{3}}{2}\right)$$

$$y = 2\sqrt{3}.$$

We then write P as $P = \left(2, -2\sqrt{3}\right)$.

LESSON 13

Example: Find the polar coordinates of $P = (2,3)$.

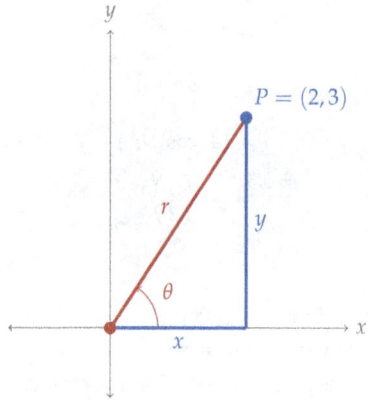

The radial coordinate r will be fairly simple to find here, via the Pythagorean Theorem:

$$r^2 = x^2 + y^2 = (2)^2 + (3)^2 = 13.$$

Therefore $r = \sqrt{13}$, and this is a positive square root as r represents the radial distance from the origin to P. Next we will determine the angle θ, via the formula

$$\tan \theta = \frac{y}{x} = \frac{3}{2}.$$

Note that this point is in the first quadrant, so we are looking for a positive, acute angle. A calculator tells us that

$$\arctan\left(\frac{3}{2}\right) \approx 56.31°.$$

Thus we'll approximate our polar coordinates for P with

$$P = \left(\sqrt{13},\, 56.31°\right).$$

Theorem: the following formulas can be used to convert between Cartesian and polar coordinate systems.

$$x = r \cos \theta \qquad r^2 = x^2 + y^2$$
$$y = r \sin \theta \qquad \tan \theta = \frac{y}{x}$$

When given r and θ, the formulas for x and y are well defined and give unique answers; that is, x and y are legitimate functions of r and θ. The formulas in the other direction, however, involve some choices. For example, when taking the square root of r in the formula

$$r^2 = x^2 + y^2,$$

we recognize that r represents a distance. Therefore we always choose the **positive** square root.

Similarly, when given x and y and finding the angular coordinate of a point, we typically choose the **least positive** angle. Choosing positive angles coincides with the convention that we measure angles counterclockwise from the positive x-axis. When we take the least such angle, this guarantees that the choice is unique, lying on the interval $\theta \in [0, 2\pi)$.

There is just one more caveat that stems from these formulas. We illustrate this with an example.

Example: Convert the point $P = (0, -2)$ from Cartesian coordinates to polar coordinates.

An image depicting this point P is shown at right. If we rely only on our formulas to convert these coordinates, finding the radial coordinate is straightforward:

$$r = \sqrt{(0)^2 + (-2)^2} = \sqrt{4} = 2.$$

But the formula $\tan \theta = \dfrac{y}{x}$ gives us multiple problems.

For one, the fraction on the right has a zero in the denominator, making it undefined. Second, in the image it is visibly obvious that the polar angle is $\theta = 270°$, but the tangent function is undefined for that angle. This makes the left-hand side of our formula meaningless as well.

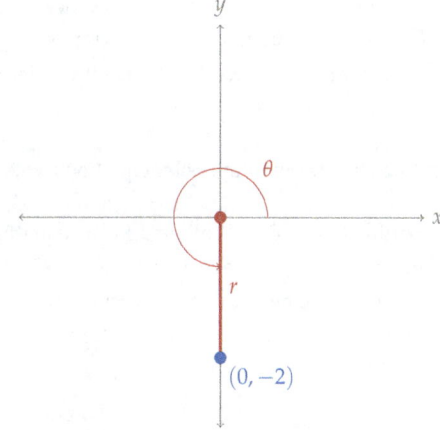

Fortunately, all such points have $x = 0$, meaning that they lie on the y-axis. Their polar angle will therefore be either $\theta = 90°$ if y is positive, or $\theta = 270°$ if y is negative.

Since our y-coordinate lies on the negative y-axis, we conclude that the polar coordinates of P are

$$P = \left(2, \frac{3\pi}{2}\right).$$

Again, the polar coordinates of this point are unique on the intervals $r \in [0, \infty)$ and $\theta \in [0, 2\pi)$.

Lesson 13

Now that we know how to convert between coordinate systems, we will move to equations and their graphs. A **polar equation** is just an expression that relates the polar coordinates r and θ.

Example: Sketch a graph of the polar equation $r = 2.75$.

Here the coordinate θ is not mentioned in the equation. It has no relation to r and may take on any value. We choose several such angles below. In each direction chosen, we label a point that is distance $r = 2.75$ away from the origin.

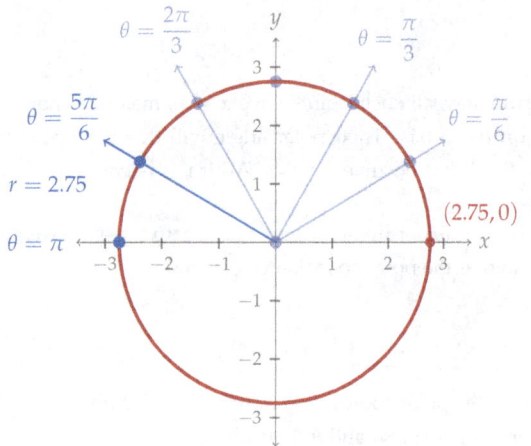

Doing this for all possible angles θ will fill in the spaces between the individual points, completing a continuous circle of radius $r = 2.75$.

Alternatively, we could convert this polar equation into a rectangular equation, involving the Cartesian coordinates x and y. We do this in the same way that we convert individual points, by using the conversion formulas.

We recall that $x^2 + y^2 = r^2$. Since we know that $r = 2.75$ for all points on our graph, we can substitute this directly into the conversion formula:

$$x^2 + y^2 = (2.75)^2.$$

This is the definition of a circle that we saw in Lesson 12, which came directly from the Pythagorean Theorem and the distance formula. It represents the set of points having distance $r = 2.75$ from the origin.

Simple equations like this are popular first examples, and we will see similar equation involving θ shortly. But, for a bit of contrast, let's first turn to an equation where r and θ **are** related. That is, both the radial coordinate r and the angular coordinate θ will appear together in the equation. The resulting graph will not be a circle, of course.

LESSON 13

Example: Sketch a graph of the polar equation $r = \sin\frac{\theta}{2}$, for θ in the interval $I = [0, 2\pi]$.

Using the same method as before, we choose several values of θ and calculate r for each choice. In the direction $\theta = 150°$, for example, we use a calculator to find that:

$$r = \sin\frac{\theta}{2} = \sin\frac{5\pi}{12} \approx 0.9659.$$

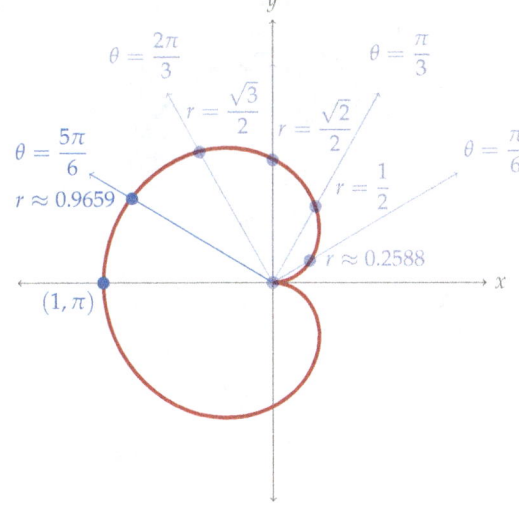

Therefore we place a point at that distance from the origin, in the chosen direction.

Continuing in this manner we find that different angles give us different radii, and only by plotting many of the angles in I can we get a clear picture of our curve. A complete graph is shown in red in the image at left.

We may decide that we want to convert this equation to rectangular coordinates as we did with the last example. However, the expression $\sin(\theta/2)$ does not fit with any of our conversion formulas. To rewrite this in terms of either $\sin\theta$ or $\cos\theta$, we search through our previous lists of trigonometric identities. We find the **half-angle formula**:

$$\sin\frac{\theta}{2} = \sqrt{\frac{1-\cos\theta}{2}}.$$

Substituting this into our given equation, we solve for $\cos\theta$:

$$r = \pm\sqrt{\frac{1-\cos\theta}{2}}$$
$$r^2 = \frac{1-\cos\theta}{2}$$
$$2r^2 = 1 - \cos\theta$$
$$\cos\theta = 1 - 2r^2.$$

149

Lesson 13

Seeing the cosine function on the left, we may wish to use the conversion $x = r\cos\theta$. However, we are still missing a factor of r. To get closer, we multiply and divide by r, and make all of the substitutions that result:

$$\frac{r\cos\theta}{r} = 1 - 2r^2$$

$$\frac{x}{\sqrt{x^2+y^2}} = 1 - 2\left(x^2+y^2\right)$$

$$\frac{x}{\sqrt{x^2+y^2}} = 1 - 2x^2 - 2y^2.$$

Note that the Cartesian form of this equation is quite a bit more complicated than its polar form. This is common for "round" graphs that are more naturally described radially.

In the previous two examples we've chosen values of θ and used the resulting radii r to plot points in the plane. In certain cases we might like to do the opposite, and choose values of r.

Each choice of radius will give us a circle of that radius. Within each circle the resulting values of θ will show us the directions at which to plot points. We show the simplest example here, in which the values of θ are constant and do not depend on r at all.

Example: Sketch a graph of the polar equation $\theta = \dfrac{5\pi}{3}$.

In this case, we are only given information about θ, which is fixed. Therefore we plot a point in this direction, $300°$, for each possible choice of r. This forms a straight line segment in that direction. We continue to assume that r measures a distance and is therefore in the interval $[0, \infty)$.

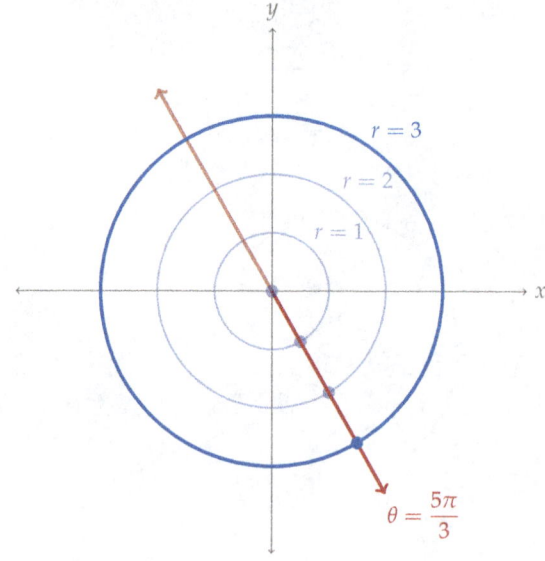

However, if we allow both positive and negative r, we can interpret each point with a "negative radius" $(-r, \theta)$ to lie opposite the point (r, θ) across the origin. This would complete the line in both directions, which we have shown in light red.

Converting this to Cartesian coordinates will require the conversion formula for θ:

$$\tan \theta = \frac{y}{x}.$$

Substituting for θ, we have

$$\frac{y}{x} = \tan \frac{5\pi}{3}$$
$$\frac{y}{x} = -\sqrt{3}$$
$$y = -\sqrt{3}x.$$

This describes a straight line of slope $m = -\sqrt{3}$, passing through the origin. Of course, in this description the radial coordinate r does not appear, so we are unable to see the extra condition that r should be positive.

As a final example, we show a graph that can be constructed with either of our strategies. In this image we choose values of θ first, but there is little reason to prefer this choice.

Example: Suppose that θ is measured in radians. Sketch a graph of the polar equation $r = \theta$.

In this equation the radial distances are equal to the angle at which they are measured. Therefore our radius grows along with our angle, constructing the following spiral shape.

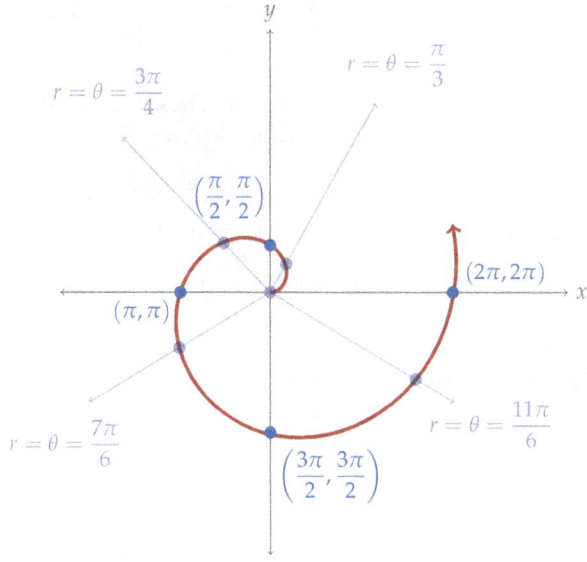

LESSON 13

EXERCISES

Plot the points whose polar coordinates are given. Then find the Cartesian coordinates of the point.

1. $\left(5, \dfrac{\pi}{2}\right)$
2. $(2, 0)$
3. $\left(2, -\dfrac{\pi}{3}\right)$
4. $(-1, \pi)$
5. $\left(7, -\dfrac{5\pi}{6}\right)$
6. $\left(-3, \dfrac{\pi}{5}\right)$
7. $\left(-2, \dfrac{7\pi}{6}\right)$
8. $(3, 2)$
9. $\left(6, -\dfrac{5\pi}{6}\right)$

The following points are written in Cartesian coordinates. Find their polar coordinates, with $r \in [0, \infty)$, $\theta \in [0, 2\pi)$.

10. $(1, 0)$
11. $(3, 3)$
12. $(0, 7)$
13. $\left(1, -\sqrt{3}\right)$
14. $(5, 12)$
15. $\left(-4\sqrt{2}, 4\sqrt{2}\right)$
16. $\left(2\sqrt{3}, -2\right)$
17. $(0, -2)$
18. $(4, 3)$

Shade the region of the plane consisting of points whose polar coordinates satisfy the given conditions.

19. $r \geq 2$
20. $\theta \in \left[0, \dfrac{\pi}{4}\right]$
21. $r \in [0, 2]$, $\theta \in \left[\dfrac{\pi}{2}, \pi\right]$
22. $r \in [1, 4]$, $\theta \in \left[\dfrac{\pi}{4}, \dfrac{\pi}{2}\right]$
23. $r \in [0, 1]$, $\theta \in \left[\dfrac{\pi}{6}, \dfrac{\pi}{3}\right]$
24. $r \in [2, 3]$, $\theta \in \left[-\dfrac{\pi}{4}, \dfrac{\pi}{4}\right]$

Find the polar form of each of the equations.

25. $y = 7$
26. $x = 2$
27. $y = x$
28. $xy = 3$
29. $x^2 + y^2 = 36$
30. $x^2 - y^2 = 16$
31. $y = x^2$
32. $x^2 - 4x + y^2 = 0$
33. $4x - 5y = 1$

Find the rectangular form of each of the equations.

34. $\theta = -\dfrac{\pi}{6}$
35. $r = 25$
36. $r = \dfrac{5}{3 - 4\sin\theta}$
37. $r = 2\sin\theta$
38. $r = \sin 2\theta$
39. $r(\sin\theta - 2\cos\theta) = 6$
40. $r = \cot\theta$
41. $r = \tan\theta$
42. $r^2 = \theta$

43. Sketch the graph of the polar equation $r = \sin(n\theta)$ for several positive integers n. How is the number of loops in each graph related to n?

44. Verify that the distance between two points $P = (r_1, \theta_1)$ and $Q = (r_2, \theta_2)$ is given by the formula

$$\mathrm{dis}(P, Q) = \sqrt{r_1^2 + r_2^2 - 2r_1 r_2 \cos(\theta_2 - \theta_1)}.$$

Use this formula to find the distance between $(3, 60°)$ and $(5, 170°)$. Round your answer to two decimal places.

LESSON 14

Cylindrical Coordinates

Now that we've established both Cartesian coordinates in three dimensions and polar coordinates in two dimensions, we can discuss a three-dimensional analogue of polar coordinates called **cylindrical coordinates**. Here we recall the conversions to polar coordinates, along with a third, trivial one for z.

$$x = r\cos\theta \qquad r^2 = x^2 + y^2$$
$$y = r\sin\theta \qquad \tan\theta = \frac{y}{x}$$
$$z = z \qquad z = z$$

The idea is to convert two of our three coordinates to polar coordinates, leaving the third alone. This conversion commonly occurs in the x- and y-variables, but this choice is not a necessary one.

Definition: Let $P = (x, y, z)$ lie in a Cartesian coordinate system, and let $P_{xy} = (x, y, 0)$ be the projection of P onto the x, y-plane. The length r of the line segment from the origin to P_{xy} is called the **radial coordinate** or **radius** of P. The angle θ created by the radius and the positive x-axis is called **angular coordinate** or **polar angle** of P.

LESSON 14

With these definitions, we then denote P in cylindrical coordinates by $P = (r, \theta, z)$, as in the image below:

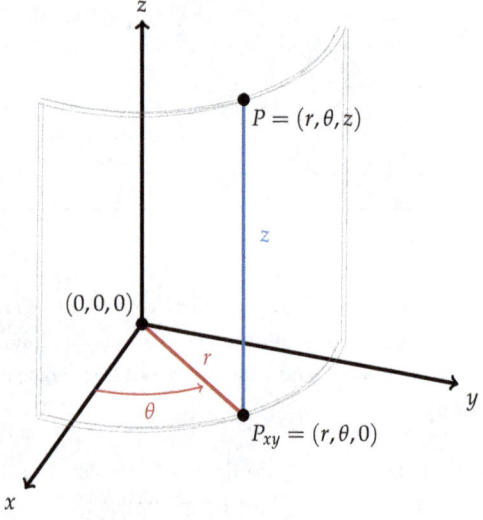

Note that the angle θ originates at the positive x-axis, and increases in the direction of the positive y-axis, just as it did for polar coordinates. This orientation is induced by the fact that we write $x = r\cos\theta$ and $y = r\sin\theta$, and not the other way around.

Example: Find the rectangular coordinates of the point

$$P = \left(2, \frac{\pi}{6}, -3\right).$$

Here we are given a point in cylindrical coordinates $P = (r, \theta, z)$, and we will use the standard conversions:

$$x = r\cos\theta = 2\cos\left(\frac{\pi}{6}\right) = 2\left(\frac{\sqrt{3}}{2}\right) = \sqrt{3},$$

$$y = r\sin\theta = 2\sin\left(\frac{\pi}{6}\right) = 2\left(\frac{1}{2}\right) = 1,$$

$$z = z = -3.$$

So in rectangular coordinates we have $P = \left(\sqrt{3}, 1, -3\right)$.

Here we see that converting a point from cylindrical to Cartesian coordinates is a relatively simple process. The method is nearly identical to the conversion from polar to Cartesian coordinates.

As in that scenario, however, converting in the other direction will require a bit more thought. Since we are asked to solve an equation for θ, we must pay attention to the **octant** in which our point lies to interpret the angles and signs correctly.

LESSON 14

Example: Find the cylindrical coordinates of the point

$$P = (3, -4, -1).$$

Keeping with our previous examples, we will convert the x- and y-values to polar coordinates.

We start with a reference triangle that shows the point's projection $P_{x,y}$ in the x,y-plane. Since P lies in Octant VIII, its projection lies in Quadrant IV of the plane.

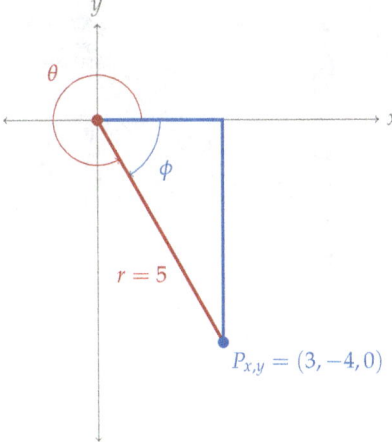

We may focus on the projection P_{xy} because the z-value of P is not being converted, and it will play no role in finding either r or θ.

To find the r-coordinate of P, we calculate:

$$r^2 = x^2 + y^2 = (3)^2 + (-4)^2 = 9 + 16 = 25.$$

Since r is a radial distance and should always be positive, we conclude that $r = 5$.

Next we will find the angular coordinate of P. Since P_{xy} lies in Quadrant IV of the x,y-plane, we should expect θ to be between $270°$ and $360°$. However, we'll first find a reference angle ϕ. A calculator tells us that

$$\phi = \arctan\left(\frac{4}{3}\right) \approx 53.13°.$$

Therefore our angle θ, measured from the positive x-axis in the direction of the positive y-axis, is approximately

$$\theta \approx 360° - 53.13° = 306.87°.$$

We then approximate the cylindrical coordinates of P as

$$P = (r, \theta, z) \approx (5, 306.87°, -1).$$

LESSON 14

Example: Sketch a graph of the equation $r = 1.4$.

We are given information about only one of the three variables. If we consider only the x,y-plane, and as we did in Lesson 13, then plotting $r = 1.4$ for various values of θ will yield a full circle.

More specifically, we might square both sides of the equation, giving $r^2 = (1.4)^2$. We can convert this to Cartesian coordinates by substituting for r^2, and obtain the equation

$$x^2 + y^2 = (1.4)^2.$$

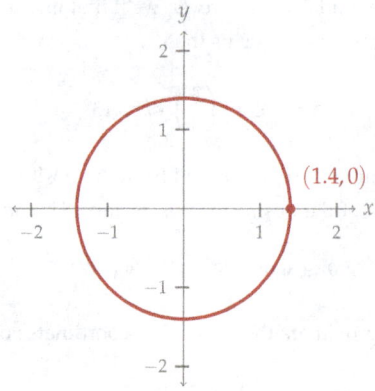

So either of these equations indicates a circle of radius 1.4 in the x,y-plane. However, this equation is meant to describe a set of points in $\mathbb{R}^3$, and no information is given about the coordinate z and its relation to r or θ.

Each point on this circle therefore defines a vertical **line** of points extending in the positive and negative z-directions. These vertical lines together form a **cylinder** in $\mathbb{R}^3$.

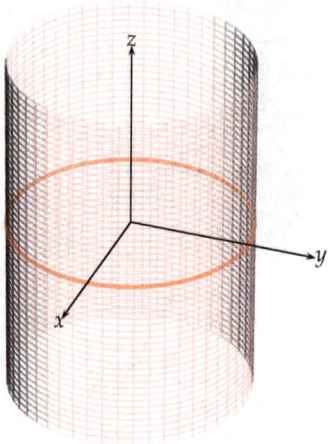

Example: Sketch a graph of the equation $\theta = \frac{\pi}{3}$.

Again, we have information about only one of our three variables. Since we are still considering x and y to be the variables changed to polar coordinates, with the standard orientation, we can draw this angle in the x,y-plane. If we extend this line infinitely in both directions, every point on it will have a θ coordinate of 60°.

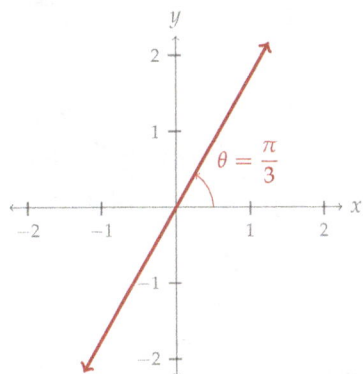

Alternatively, we might again convert this equation to rectangular coordinates by taking the tangent of each side:

$$\frac{y}{x} = \tan \theta = \tan\left(\frac{\pi}{3}\right) = \sqrt{3}.$$

This gives us the Cartesian equation $y = \sqrt{3}x$, which is indeed the line shown at left.

Since we also have no information about z, each point on this line will itself generate a vertical line of points extending in the positive and negative z-directions. The set of these vertical lines together forms a **vertical plane** in $\mathbb{R}^3$.

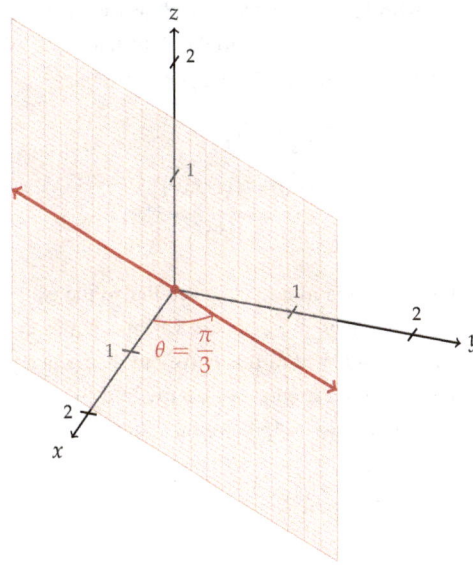

LESSON 14

Example: Use cylindrical coordinates to identify and sketch a graph of the following equation in $\mathbb{R}^3$:

$$x^2 + y^2 - 2y = 0.$$

We immediately notice that the variable z is not present in this equation, so we know that if we graph this function in the x, y-plane, each point on that graph will extend in a vertical line, creating a surface. We will simplify this equation first by moving the linear term to the right-hand side of the equation, then applying our conversion formulas.

$$x^2 + y^2 = 2y$$
$$(r^2) = 2(r \sin \theta)$$
$$r = 2 \sin \theta.$$

We can now treat this in a manner similar to the examples in Lesson 13. Namely, if we choose various angles θ, each will determine a different radius r. We can then plot a point that distance from the origin, along a ray in the direction of θ. For example, when $\theta = 150°$ we calculate that

$$r = 2 \sin \left(\frac{5\pi}{6} \right) = 2 \left(\frac{1}{2} \right) = 1.$$

We plot several such points to create the following graph.

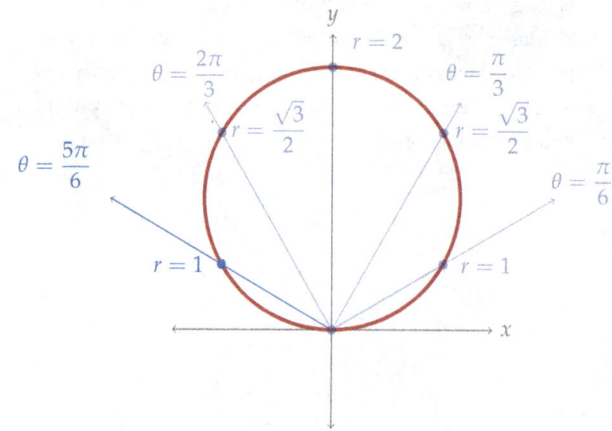

Since there is no restriction on the values of θ, we might continue plotting points for larger angles. If we do this, angles in Quadrants III and IV will give us values of r that are negative:

$$r = 2 \sin \left(\frac{7\pi}{6} \right) = 2 \left(-\frac{1}{2} \right) = -1.$$

Again, we interpret this to mean that on the ray pointing in the $\theta = 210°$ direction, our point lies 1 unit from the origin, in the direction **opposite** to the way the ray points.

In this case these points continue to lie on the same circle that we have already generated. Therefore allowing θ to increase further will simply generate the same circle many times. This forms the complete graph of $x^2 + y^2 - 2y = 0$ in the x, y-plane.

Again, this equation is intended to describe a set of points in $\mathbb{R}^3$, so this graph should be extended into three dimensions. With no information about z and its relationship to either r or θ, each of these points on the circle extends vertically into a line.

The resulting graph in $\mathbb{R}^3$ is a **circular cylinder**. Its axis of symmetry lies along the y-axis, and the cylinder itself lies tangent to the z-axis.

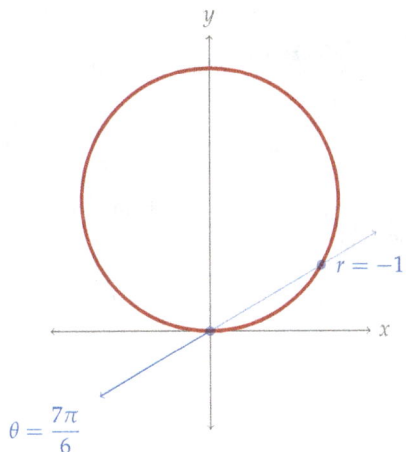

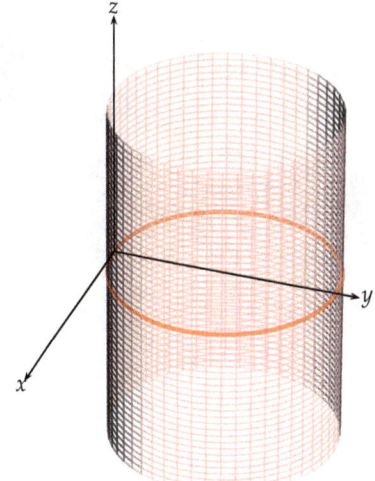

Lesson 14

Example: Sketch a graph of the equation $z = 1.3$.

In the case that the only specified variable is the one remaining in Cartesian coordinates, we can simply treat the equation as if it actually were written in Cartesian coordinates. In this case, the set of points satisfying this equation are those whose z-coordinate is equal to 1.3, and the coordinates in which x and y are written are irrelevant.

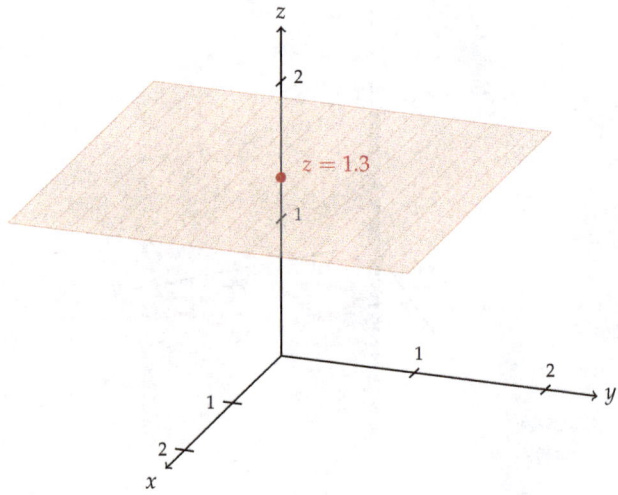

Throughout this series of examples we have been using the standard definition for cylindrical coordinates. However, there are several other choices of coordinates that could rightly be called "cylindrical." These can be achieved by slightly altering the standard conversion formulas. Let's explore one alternate option and its orientation.

For this orientation we will elect to change x and z to polar coordinates, leaving y alone. More specifically, we will set $x = r\cos\theta$, and $z = r\sin\theta$, resulting in the following set of conversions:

$$x = r\cos\theta \qquad r^2 = x^2 + z^2$$
$$y = y \qquad y = y$$
$$z = r\sin\theta \qquad \tan\theta = \frac{z}{x}$$

Since the x is associated to cosine here, it is easy to check that the following cylindrical point,

$$P = (r, y, \theta) = (1, 0, 0°),$$

lies at the Cartesian point $P = (x, y, z) = (1, 0, 0)$ as well. However, the cylindrical point

$$Q = (r, y, \theta) = (1, 0, 90°)$$

lies at the Cartesian point $Q = (x,y,z) = (0,0,1)$, showing that the angle θ originates at the positive x-axis and increases in the direction of the positive z-axis. The y-coordinate, of course, measures the distance of a point from the x,z-plane. We can see the placement of a general point in the image below.

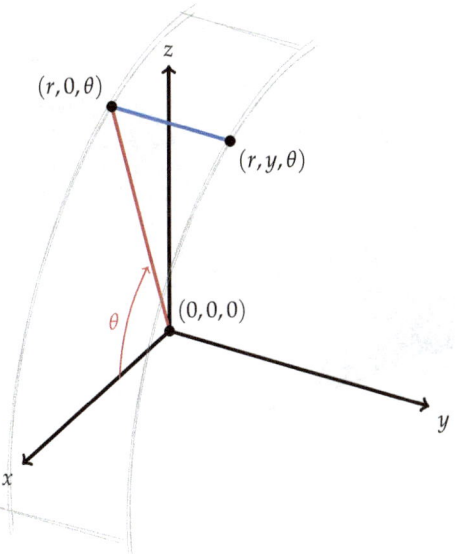

With this alternate change of coordinates, we find that an equation of the form $r = 1.4$ still generates a cylinder in space, as it did before. However, here it would be oriented with its central axis of symmetry aligned with the y-axis. Also, the corresponding Cartesian equation would be

$$r = 1.4$$
$$r^2 = (1.4)^2$$
$$x^2 + z^2 = (1.4)^2.$$

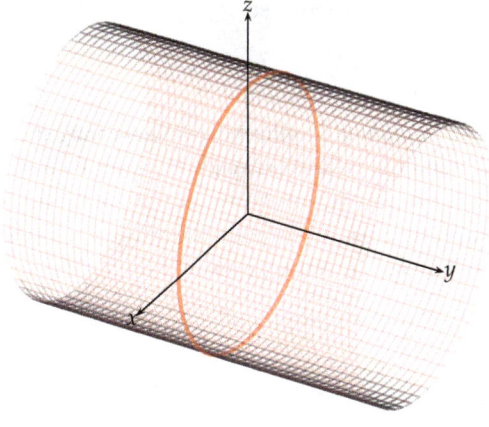

LESSON 14

As a more complex example, consider the equation $r = \theta$. In Lesson 13 we saw that this forms a spiral in the plane.

However, in these current coordinates the rotation of θ occurs from the positive x-axis toward the positive z-axis, so this spiral will occur in the x, z-plane. Each point on the spiral generates a line of points extending in the y-direction. For $\theta \in [0, 2\pi]$, we obtain the surface shown at right.

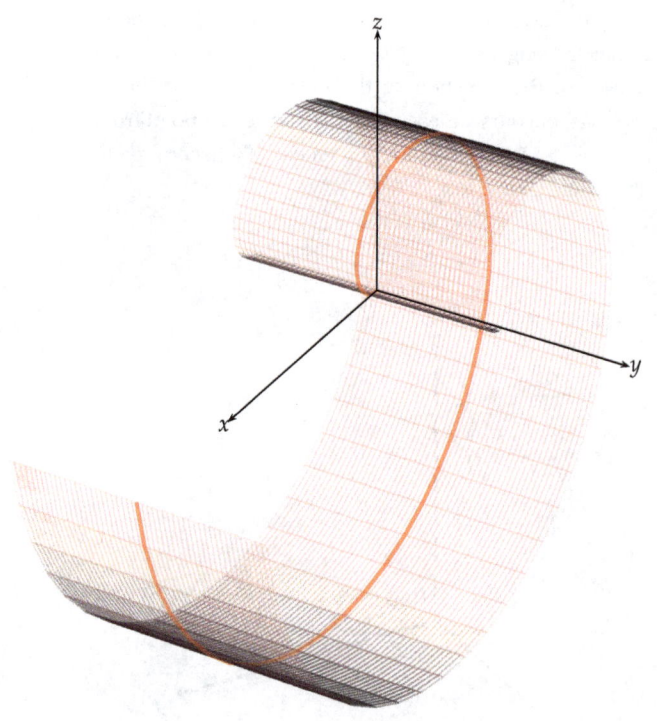

It should now be clear that Cartesian coordinates can be converted to cylindrical coordinates in many ways, depending on the orientation you would like the coordinate system to have. With an equation like

$$y^2 + z^2 = 9,$$

we would be likely to do something different altogether and change y and z to polar coordinates. However, if no specification is given you should always assume that the standard conversion formulas are in place. Once again, those standard conversions are written below.

$$\begin{aligned} x &= r\cos\theta & r^2 &= x^2 + y^2 \\ y &= r\sin\theta & \tan\theta &= \frac{y}{x} \\ z &= z & z &= z \end{aligned}$$

LESSON 14

EXERCISES

The following points are given in cylindrical coordinates in x and y. Convert each point to Cartesian coordinates.

1. $\left(2, \frac{\pi}{3}, 2\right)$
2. $\left(2, \frac{4\pi}{3}, 8\right)$
3. $\left(3, \frac{\pi}{2}, 1\right)$
4. $\left(4, \frac{7\pi}{6}, 3\right)$
5. $(5, 0, 2)$
6. $\left(\sqrt{2}, \frac{\pi}{4}, \sqrt{2}\right)$
7. $(1, \pi, e)$
8. $\left(1, \frac{3\pi}{2}, 1\right)$
9. $(0, \pi, 7)$

Convert the points from Cartesian coordinates to cylindrical coordinates in x and y, with $r \in [0, \infty)$ and $\theta \in [0, 2\pi)$.

10. $(0, 3, 5)$
11. $(-1, 0, 0)$
12. $(1, 1, 1)$
13. $\left(1, \sqrt{3}, 4\right)$
14. $(-3, 2, -1)$
15. $\left(-\sqrt{2}, \sqrt{2}, 0\right)$
16. $\left(\sqrt{3}, 1, 9\right)$
17. $(2, -2, 4)$
18. $(0, 0, -1)$

Find the cylindrical form of each of the equations. Make the conversions in the variables x and y.

19. $x^2 + y^2 + z^2 = 25$
20. $x^2 + y^2 - z^2 = 25$
21. $9x^2 + 9y^2 = z^2$
22. $y = 4$
23. $x^2 + y^2 = 4y$
24. $x^2 - y^2 - 2z^2 = 4$

Match each of the following equations with its graph.

25. $r = 2$
26. $z = r^2$
27. $r^2 + z^2 = 2.25$
28. $r = z$

(a)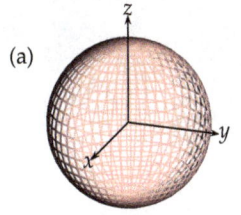

(b)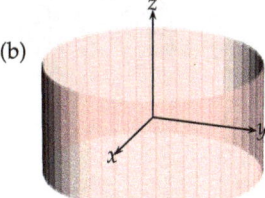

(c)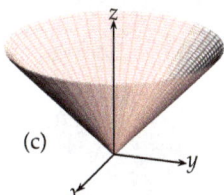

(d)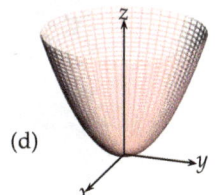

29. Verify that the distance between $P = (r_1, \theta_1, z_1)$ and $Q = (r_2, \theta_2, z_2)$ in $\mathbb{R}^3$ is given by the formula

$$\text{dis}(P, Q) = \sqrt{z_1^2 + z_2^2 + r_1^2 + r_2^2 - 2z_1 z_2 - 2r_1 r_2 \cos(\theta_2 - \theta_1)}.$$

Use this formula to find the distance between $(4, 120°, -1)$ and $(3, 210°, 5)$. Round your answer to two decimal places.

LESSON 15

Spherical Coordinates

If you have ever studied a globe or a map of the Earth, you are likely to be familiar with the concepts of **latitude** and **longitude**. This grid-like system allows us to specify points on the surface of the Earth in terms of two **angles**, each of which has its vertex at the center of the Earth.

One of these angles, the latitude, is measured either north or south from the equator. The longitude is measured from a circle that passes through both poles and the city of Greenwich, England, which was chosen due to its importance in maritime history. The "origin" of this coordinate system thus lies in the Gulf of Guinea, just south of Accra, Ghana.

Here we will adapt this concept to create a new coordinate system for $\mathbb{R}^3$, which we call **spherical coordinates**. However, we will need to make a few key modifications.

For one, we will also need to include a third coordinate for this three-dimensional coordinate system. Unlike points on the two-dimensional surface of the Earth, points in space may have varying distances from the center. So in some sense, you could think of our additional third coordinate as measuring the **altitude** of a point, measured from the center of the Earth. This implies that the origin of our coordinate system will be this same center.

LESSON 15

The other main modification is to the angle that corresponds to latitude, which will now be measured from the "north pole" instead of from the "equator." We'll be more precise with these definitions below. We begin with an already established rectangular coordinate system, and our new system will be constructed from it.

Definition: Let $P = (x, y, z)$ be a point in a three-dimensional Cartesian coordinate system, and let $P_{xy} = (x, y, 0)$ be the projection of P onto the x, y-plane.

- The distance r from the origin to P_{xy} is called the **polar radius** of P.

- The angle θ created by the polar radius and the positive x-axis is called the **polar angle** of P.

- The distance ρ from the origin to the point P itself is called the **spherical radius** of P.

- The angle ϕ created by the spherical radius and the positive z-axis is called the **azimuth angle** of P.

Since this is a three-dimensional space, we should need only three pieces of information to uniquely determine a point, and indeed this is the case here. Generally speaking the polar radius r is an unnecessary part of the spherical coordinate system. We mention it here to distinguish it from the spherical radius ρ, and because we will use it in an upcoming calculation.

In the image below, we see a point P in a coordinate system, along with its projection P_{xy}.

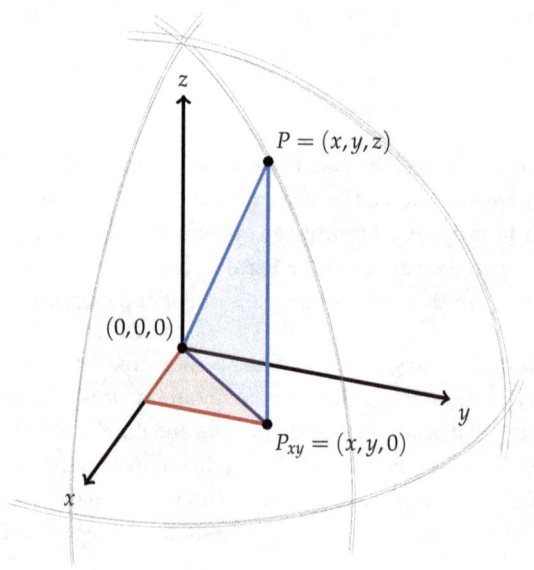

The two colored triangles are drawn independently in their own two-dimensional coordinate systems below. One lies entirely in the x, y-plane, while the other is formed by the origin, P, and P_{xy}. The polar and spherical radii, the polar angle, and the azimuth angle are all labeled appropriately in that image.

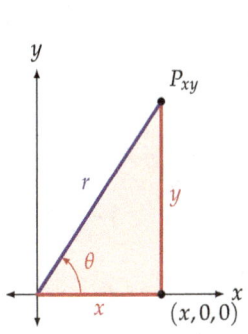 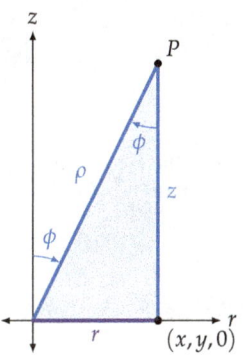

Examining first the red triangle in the x, y-plane, we find that it satisfies the standard formulas for polar coordinates. That is, we know that:

$$x = r \cos \theta \qquad r^2 = x^2 + y^2$$
$$y = r \sin \theta \qquad \tan \theta = \frac{y}{x}$$

Turning to the blue triangle, we use the definitions of the trigonometric functions to see that

$$r = \rho \sin \phi,$$
$$z = \rho \cos \phi.$$

To write the coordinates x and y entirely in terms of ρ, θ, and ϕ, we combine these sets of expressions.

$$x = r \cos \theta \qquad y = r \sin \theta$$
$$x = (\rho \sin \phi) \cos \theta \qquad y = (\rho \sin \phi) \sin \theta$$

Another connection between these coordinates can be found using the Pythagorean Theorem. From the red and blue triangles, respectively, we know that

$$r^2 = x^2 + y^2 \quad \text{and} \quad \rho^2 = r^2 + z^2.$$

Combining these, we find that

$$\rho^2 = (x^2 + y^2) + z^2.$$

In this way we recover the same formula for the distance to the origin that we saw in Lesson 12. To find the distance between two points away from the origin, we could draw a similar picture, again with two right triangles, and repeat this argument.

Lesson 15

We summarize our new conversion formulas here. The last formula is a direct rearrangement of the other expressions.

$$x = \rho \sin\phi \cos\theta \qquad \rho^2 = x^2 + y^2 + z^2$$

$$y = \rho \sin\phi \sin\theta \qquad \tan\theta = \frac{y}{x}$$

$$z = \rho \cos\phi \qquad \cos\phi = \frac{z}{\rho} = \frac{z}{\sqrt{x^2+y^2+z^2}}$$

Example: Find the Cartesian coordinates of the point

$$P = \left(8, -\frac{5\pi}{6}, \frac{\pi}{4}\right).$$

The point is given to us in spherical coordinates $P = (\rho, \theta, \phi)$, so we use the standard formulas, finding first the x-coordinate:

$$x = \rho \sin\phi \cos\theta$$

$$= 8 \sin\left(\frac{\pi}{4}\right) \cos\left(-\frac{5\pi}{6}\right)$$

$$= 8 \left(\frac{\sqrt{2}}{2}\right) \left(-\frac{\sqrt{3}}{2}\right)$$

$$= -2\sqrt{6}.$$

Next we find the y-coordinate:

$$y = \rho \sin\phi \sin\theta$$

$$= 8 \sin\left(\frac{\pi}{4}\right) \sin\left(-\frac{5\pi}{6}\right)$$

$$= 8 \left(\frac{\sqrt{2}}{2}\right) \left(-\frac{1}{2}\right)$$

$$= -2\sqrt{2}.$$

Finally, the z-coordinate is

$$z = \rho \cos\phi = 8\cos\left(\frac{\pi}{4}\right) = 8\left(\frac{\sqrt{2}}{2}\right) = 4\sqrt{2}.$$

Therefore, P is written in Cartesian coordinates as

$$P = \left(-2\sqrt{6}, -2\sqrt{2}, 4\sqrt{2}\right).$$

In this book we will try to be consistent in the way that we order the angles θ and ϕ in these spherical points, with θ being listed first. However, we should be warned that authors of other books or articles may order these angles differently. It is always a good idea to double check the convention being used.

As with polar and cylindrical coordinates, the spherical representation of a point will not be unique. For example, the spherical coordinates

$$\left(5, \frac{\pi}{2}, -\frac{5\pi}{3}\right) \quad \text{and} \quad \left(5, \frac{5\pi}{2}, \frac{\pi}{3}\right)$$

actually both represent the same point in space. However, it is customary to introduce a limited amount of uniqueness by insisting that the values of each coordinate lie within the following intervals:

$$r \in [0, \infty),$$
$$\theta \in [0, 2\pi),$$
$$\phi \in [0, \pi].$$

Every point in space will have a spherical coordinate representation within these bounds. Note the bounds for ϕ, while recalling again that it is measured downward from the positive z-axis. Points with a positive z-coordinate can be represented by a ϕ-coordinate between $0°$ and $90°$, while points with negative z-height can be represented by a ϕ in the interval $[90°, 180°]$. Therefore there is no need to multiply define a point by choosing a ϕ larger than π. This interval also coincides with the fundamental domain of cosine, which is useful in the next example.

Example: Find the spherical coordinates of the point

$$P = (2, -3, 5).$$

We will find each spherical coordinate in turn, starting with the simplest to find, the spherical radius ρ. For this we do the conversion based on the Pythagorean Theorem:

$$\rho^2 = x^2 + y^2 + z^2 = (2)^2 + (-3)^2 + (5)^2 = 38.$$

As usual, we choose the positive square root, so $\rho = \sqrt{38}$.

Next we find the azimuth angle ϕ. With a positive z-coordinate, we see that P lies above the x,y-plane, so we expect ϕ to be in the interval $[0°, 90°]$. We use the formula $z = \rho \cos \phi$ here:

$$z = \rho \cos \phi$$
$$(5) = \sqrt{38} \cos \phi$$
$$\cos \phi = \frac{5}{\sqrt{38}}$$
$$\phi = \arccos\left(\frac{5}{\sqrt{38}}\right)$$
$$\phi \approx 35.8°.$$

Lesson 15

Finally, we find the polar angle θ in a similar fashion. For this, the projection P_{xy} is in Quadrant IV of the x,y-plane, so we expect that θ will be in the interval $[270°, 360°]$. We start with the conversion:

$$\tan \theta = \frac{y}{x}$$

$$\tan \theta = \frac{-3}{2}$$

$$\theta = \arctan\left(-\frac{3}{2}\right).$$

Now, tangent is an odd function, and therefore so is arctangent. We use this fact here:

$$\theta = -\arctan\left(\frac{3}{2}\right)$$

$$\theta \approx -56.31°.$$

To make θ positive so that it lies in the interval $[270°, 360°]$, we instead choose the coterminal angle

$$\theta \approx 360° - 56.31° = 303.69°.$$

So, we can approximate P, in spherical coordinates, by

$$P = (\rho, \theta, \phi) \approx \left(\sqrt{38}, 303.69°, 35.8°\right).$$

Example: Sketch a graph of the spherical equation $\rho = 3.2$.

In this example we are given information only about the spherical radius ρ. The set of points that are a distance of $\rho = 3.2$ forms a sphere in $\mathbb{R}^3$. We can further identify this sphere by squaring both sides of the equation, and using the conversion

$$\rho^2 = x^2 + y^2 + z^2 = (3.2)^2.$$

Of course, this is the standard form for a sphere with its origin at the center, which we saw in Lesson 12.

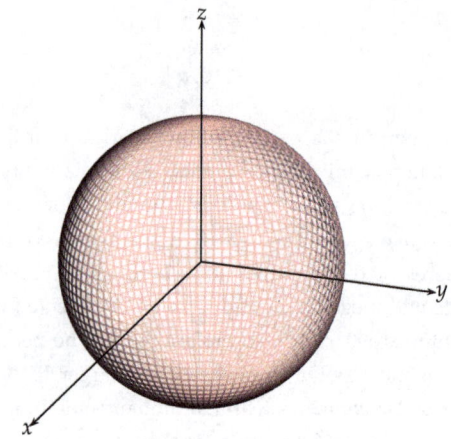

170

LESSON 15

Example: Sketch a graph of the spherical equation $\phi = \frac{2\pi}{13}$.

This equation specifies only the azimuth angle ϕ, which is measured downward from the positive z-axis, in the direction of the x, y-plane. The combination of ϕ with any fixed polar angle θ will specify a ray of points extending from the origin in the given direction. For example, when $\theta = 90°$ we obtain the ray A indicated below. For graphing purposes, note that $\phi = 27.69°$.

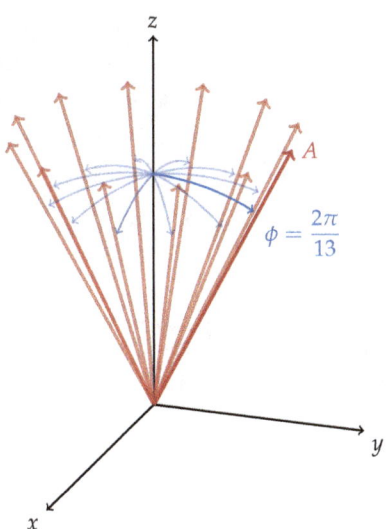

Defining a specific ϕ in this way is akin to specifying a line of latitude on a globe, but not a longitude or an altitude. Further choosing a value for θ would be enough to specify a point on a globe, but not the altitude of an object above or below that point.

Filling in all possible choices of θ on the interval $[0, 2\pi]$ completes our surface. We can see that this equation defines a **cone**, whose axis of symmetry is the positive z-axis.

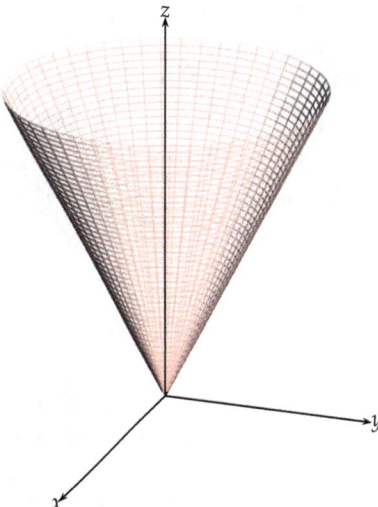

LESSON 15

Example: Sketch a graph of the equation $\theta = -\frac{\pi}{6}$.

Since θ represents the same polar angle in both cylindrical and spherical coordinates, this equation has the same graph in both systems. The solution to this example will therefore be analogous to the one from our last lesson. We again draw this angle first in the x, y-plane, and then extend it in the z-direction.

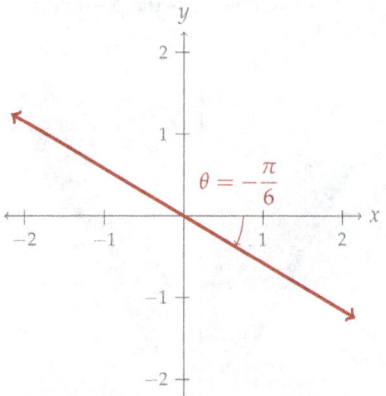

We might again convert this equation to rectangular coordinates by taking the tangent of each side of this equation:

$$\tan \theta = \tan\left(-\frac{\pi}{6}\right) = -\frac{\sqrt{3}}{3} = \frac{y}{x}.$$

This gives us the Cartesian equation $y = -\frac{\sqrt{3}}{3}x$, which is indeed the line shown at left.

Each point on this line will itself generate a vertical line of points extending in the positive and negative z-directions. The set of these vertical lines together forms a vertical plane in $\mathbb{R}^3$. Note that to obtain the full plane, as opposed to a half-plane, we must allow for either negative values of ρ or negative angles ϕ.

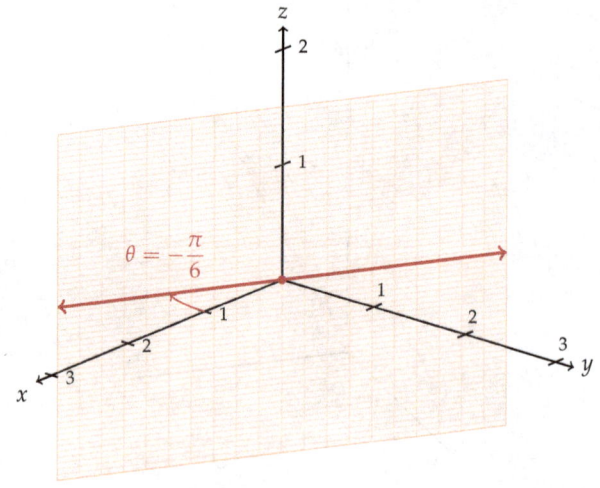

LESSON 15

EXERCISES

The following points are given in spherical coordinates (ρ, θ, ϕ). Convert each point to Cartesian coordinates.

1. $\left(4, \dfrac{\pi}{6}, \dfrac{\pi}{4}\right)$
2. $(1, 0, 0)$
3. $\left(1, \dfrac{\pi}{6}, \dfrac{\pi}{6}\right)$
4. $\left(12, \dfrac{3\pi}{4}, \dfrac{\pi}{9}\right)$
5. $(3, 0, \pi)$
6. $\left(2, \dfrac{\pi}{4}, \dfrac{\pi}{4}\right)$
7. $\left(5, \dfrac{\pi}{4}, \dfrac{3\pi}{4}\right)$
8. $\left(6, \pi, \dfrac{\pi}{2}\right)$
9. $\left(12, \dfrac{\pi}{4}, 0\right)$

Convert the points from Cartesian coordinates to spherical coordinates, with $\rho \in [0, \infty)$, $\theta \in [0, 2\pi)$, and $\phi \in [0, \pi)$.

10. $(4, 0, 0)$
11. $\left(\sqrt{3}, 0, 1\right)$
12. $(1, 1, 1)$
13. $\left(-2, 2\sqrt{3}, 4\right)$
14. $(0, -3, 0)$
15. $\left(1, -1, \sqrt{2}\right)$
16. $\left(-\sqrt{3}, -3, -2\right)$
17. $\left(\sqrt{3}, 1, 2\sqrt{3}\right)$
18. $(0, 0, 2)$

Find the spherical form of each of the equations.

19. $x^2 + y^2 + z^2 = 25$
20. $x^2 + y^2 - z^2 = 25$
21. $9x^2 + 9y^2 = z^2$
22. $y = 4$
23. $x^2 + y^2 = 4y$
24. $x^2 - y^2 - 2z^2 = 4$

Sketch the solid described by the given conditions.

25. $\rho \in [0, 3]$, $\phi \in \left[0, \dfrac{\pi}{3}\right]$
26. $r \in [0, 2]$, $\theta \in \left[0, \dfrac{\pi}{2}\right]$, $z \in [0, 4]$
27. $\theta \in \left[\dfrac{3\pi}{2}, 2\pi\right]$, $r \leq z \leq 2$
28. $\rho \in [0, 1]$, $\theta \in [0, \pi]$, $\phi \in \left[\dfrac{\pi}{4}, \dfrac{\pi}{2}\right]$

In Exercises 29–34, convert each point from cylindrical to spherical coordinates, or vice-versa.

29. $(\rho, \theta, \phi) = \left(10, \dfrac{\pi}{6}, \dfrac{\pi}{2}\right)$
30. $(r, \theta, z) = \left(4, \dfrac{\pi}{4}, 0\right)$
31. $(r, \theta, z) = \left(1, \dfrac{\pi}{2}, 1\right)$
32. $(\rho, \theta, \phi) = (5, 0, 0)$
33. $(\rho, \theta, \phi) = \left(2\sqrt{2}, \dfrac{3\pi}{2}, \dfrac{\pi}{2}\right)$
34. $(r, \theta, z) = \left(4, \dfrac{7\pi}{6}, 4\sqrt{3}\right)$

35. The following points in $\mathbb{R}^3$ are given in spherical coordinates (ρ, θ, ϕ). Find the distance between the points.

$$P = \left(10, \dfrac{3\pi}{4}, \dfrac{\pi}{4}\right) \quad Q = \left(8, \dfrac{7\pi}{4}, \dfrac{\pi}{6}\right)$$

LESSON 16

Transformations of Graphs

Imagine that we have an equation that relates two variables, which each represent numerical values written with a certain set of units. This equation will have a graph associated to it. If we wished to change the units of one of the variables, say from meters to centimeters, we would do this by multiplying this variable's values by an appropriate conversion factor.

Of course, we would then need to re-draw the graph of our modified equation. This is just one example of the **transformation** of a graph. We will explore this and several other transformations in this lesson.

Since we may be encountering these transformations for the first time, for simplicity we'll focus here on two-dimensional graphs of equations involving x and y. However, these rules are also valid for graphs in three dimensions that involve three variables, or more generally graphs in any number of variables.

We already have some experience graphing equations from the last few lessons, and we could draw from any our previous examples to illustrate graph transformations. But we'll instead focus on a new graph, which comes from an old function: **sine**.

LESSON 16

We'll construct the graph of the **sine function** by direct computation. We will need a good number of points to be able to really see what the graph looks like, so below we've made a list of values for all of the standard angles of the Unit Circle.

$$\sin(0) = 0$$
$$\sin\left(\frac{\pi}{6}\right) = \frac{1}{2}$$
$$\sin\left(\frac{\pi}{4}\right) = \frac{\sqrt{2}}{2}$$
$$\sin\left(\frac{\pi}{3}\right) = \frac{\sqrt{3}}{2}$$
$$\sin\left(\frac{\pi}{2}\right) = 1$$
$$\sin\left(\frac{2\pi}{3}\right) = \frac{\sqrt{3}}{2}$$
$$\sin\left(\frac{3\pi}{4}\right) = \frac{\sqrt{2}}{2}$$
$$\sin\left(\frac{5\pi}{6}\right) = \frac{1}{2}$$

$$\sin(\pi) = 0$$
$$\sin\left(\frac{7\pi}{6}\right) = -\frac{1}{2}$$
$$\sin\left(\frac{5\pi}{4}\right) = -\frac{\sqrt{2}}{2}$$
$$\sin\left(\frac{4\pi}{3}\right) = -\frac{\sqrt{3}}{2}$$
$$\sin\left(\frac{3\pi}{2}\right) = -1$$
$$\sin\left(\frac{5\pi}{3}\right) = -\frac{\sqrt{3}}{2}$$
$$\sin\left(\frac{7\pi}{4}\right) = -\frac{\sqrt{2}}{2}$$
$$\sin\left(\frac{11\pi}{6}\right) = -\frac{1}{2}$$

Plotting these points in the x, y-plane, we obtain a good image of the graph. The complete, smooth curve is also filled in between the individual points. We should note that for clarity, the x and y-axes are not on the same scale in these images.

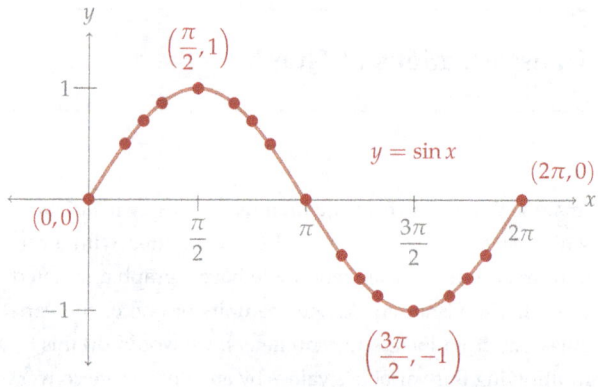

Now, we know that we can continue to evaluate the sine function for angles that are larger than $x = 2\pi$, using the Unit Circle, and we also know that these values we've found will repeat over coterminal angles. Therefore, over the interval $[2\pi, 4\pi]$ the graph will look identical to this segment over $[0, 2\pi]$.

The length of these intervals, past which sine begins to repeat its values, is called the **period** of the sine function.

We can include negative angles of x as well, by continuing to use the Unit Circle to find coterminal angles. A more complete graph of $y = \sin x$ is shown below. Again, the graph will repeat its values with period 2π infinitely many times in both directions.

We will not do so here, but evaluating the **cosine function** at all of these same angles of the Unit Circle will give us the same collection of values as sine, albeit in a different arrangement. Plotting the resulting points in the x, y-plane will give us the graph shown in blue below. We see that the graph of cosine also has period 2π. In fact, it looks identical to the graph of sine in shape, aside from the fact that it has been **shifted** or **translated** by 90° to the left.

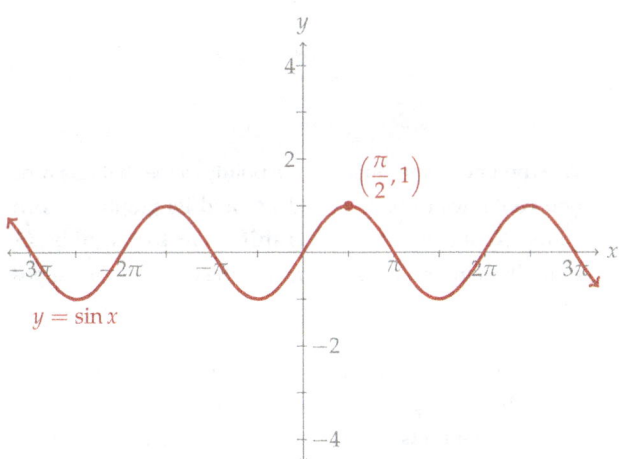

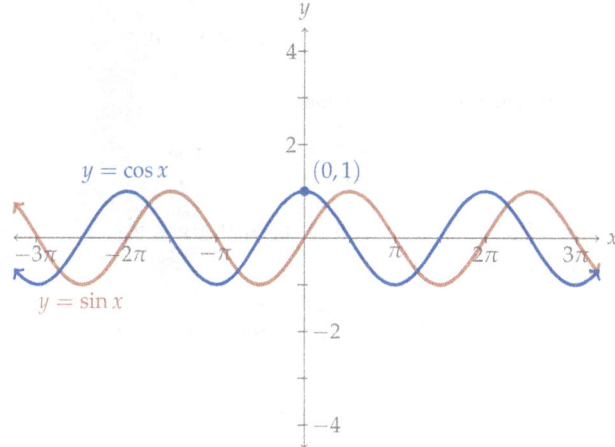

Lesson 16

This type of shift represents the first of the transformations that we'll discuss.

Example: Sketch the graph of the function

$$y = \sin\left(x - \frac{\pi}{4}\right).$$

We will plot a few representative points on this graph and use our knowledge of the standard sine graph to fill in the rest. First, when $x = 45°$, we have

$$y = \sin\left(\frac{\pi}{4} - \frac{\pi}{4}\right) = \sin(0) = 0.$$

Next, when $x = 135°$, we have

$$y = \sin\left(\frac{3\pi}{4} - \frac{\pi}{4}\right) = \sin\left(\frac{\pi}{2}\right) = 1.$$

We can also plug in each of the angles $x = 225°$ and $x = 315°$ to obtain the values

$$y = \sin\left(\frac{5\pi}{4} - \frac{\pi}{4}\right) = \sin(\pi) = 0,$$

$$y = \sin\left(\frac{7\pi}{4} - \frac{\pi}{4}\right) = \sin\left(\frac{3\pi}{2}\right) = -1.$$

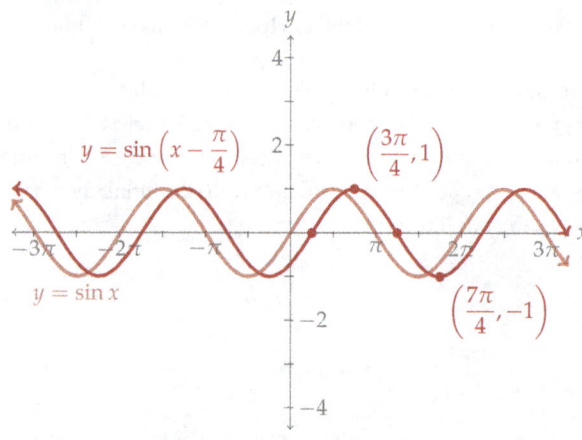

We won't need very many more points to see that this function continues to have period 2π, and its graph is shown above. Note that this graph is **shifted to the right** by 45° from the standard graph of sine, in the direction of the positive x-axis.

This will be a general rule. Replacing the horizontal variable x by the expression $(x - h)$ will shift the graph horizontally by h units. The direction of the shift is determined by the sign of h; a positive h will shift the original graph in the positive direction, and conversely for a negative h.

LESSON 16

Earlier we noticed that the graphs of sine and cosine are identical, up to a shift of 90°. More specifically, if we start with the graph of $y = \cos x$, and shift it to the right by 90° by replacing x with the expression $(x - 90°)$, we will obtain the graph of $y = \sin x$. That is,

$$\sin x = \cos(x - 90°).$$

However, you may notice that this seems to be at odds with the **Complementary Angle Theorem** that we learned earlier, which states that

$$\sin x = \cos(90° - x).$$

In fact, both of these statements are true. Cosine is an **even** function, so we can write the expression in either order:

$$\begin{aligned} \sin x &= \cos(90° - x) \\ &= \cos(-(90° - x)) \\ &= \cos(x - 90°). \end{aligned}$$

Through this relationship, we can write any graph of cosine in terms of the sine function. It is for this reason that all such graphs are simply called **sinusoidal curves**. These curves effectively model a wide variety of natural phenomena, from swinging pendulums to electromagnetic waves.

Example: Sketch the graph of the function

$$y = \sin\left(x + \frac{5\pi}{6}\right).$$

This is an equation of the form $y = \sin(x - h)$, specifically,

$$y = \sin\left(x - \left(-\frac{5\pi}{6}\right)\right).$$

Therefore this represents the standard sine graph, shifted horizontally 150° to the left. In particular, the point $(90°, 1)$ on the graph of sine is shifted to the point $(-60°, 1)$:

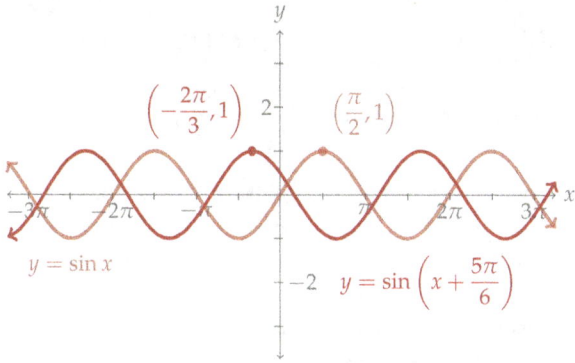

179

LESSON 16

Not surprisingly, we can shift these graphs in the **vertical** direction in a completely analogous way. This time we will replace the variable y, which may occur on either side of the equation.

Example: Sketch the graph of the function

$$y = \sin x + 2.$$

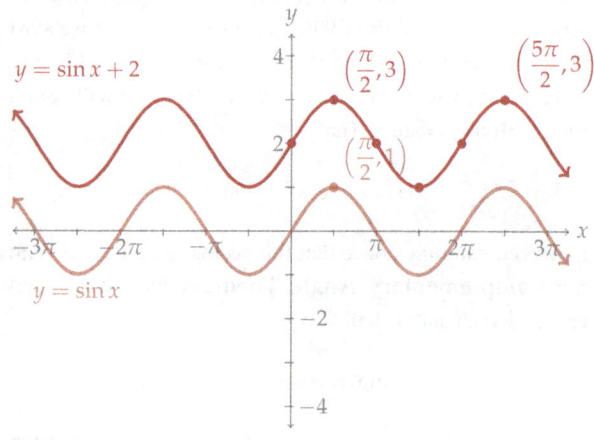

We can again plot some representative points for this graph, as we did in the previous example.

$$\sin(0) + 2 = 0 + 2 = 2 \qquad \sin\left(\frac{3\pi}{2}\right) + 2 = -1 + 2 = 1$$

$$\sin\left(\frac{\pi}{2}\right) + 2 = 1 + 2 = 3 \qquad \sin(2\pi) + 2 = 0 + 2 = 2$$

$$\sin(\pi) + 2 = 0 + 2 = 2 \qquad \sin\left(\frac{5\pi}{2}\right) + 2 = 1 + 2 = 3$$

We find the same graph as $y = \sin x$, though the constant $k = 2$ has been added to every y-coordinate, shifting the graph 2 units in the positive y-direction. To maintain consistency with what we saw in the horizontal direction, we can move the constant $k = 2$ to the other side of the equation, leaving $(y - 2) = \sin x$.

This shows that replacing y with the expression $(y - k)$ has the same effect as the replacement of the x-variable, only in the vertical, y-direction. To summarize, we list the following rules:

- Replacing x by $(x - h)$ **shifts** the graph **horizontally** by h units, right if $h > 0$, and left if $h < 0$.

- Replacing y by $(y - k)$ **shifts** the graph **vertically** by k units, up if $k > 0$, and down if $k < 0$.

Now that we know how to shift graphs both horizontally and vertically, we'll move on to another type of transformation, called a **scaling**. This time, instead of adding/subtracting a constant to a variable, we will act on the variable by **multiplication**.

Example: Sketch the graph of the function

$$y = 3\sin x.$$

We'll construct another table of values, this time by multiplying the standard values of sine by a factor of 3:

$$3\sin(0) = 3(0) = 0 \qquad 3\sin\left(\frac{3\pi}{2}\right) = 3(-1) = -3$$

$$3\sin\left(\frac{\pi}{2}\right) = 3(1) = 3 \qquad 3\sin(2\pi) = 3(0) = 0$$

$$3\sin(\pi) = 3(0) = 0 \qquad 3\sin\left(\frac{5\pi}{2}\right) = 3(1) = 3$$

Plotting these points in the image at right, we see that this modification to the equation of sine increases the vertical distance between each point and the x-axis by a factor of 3. This effectively **stretches** the graph in the vertical direction.

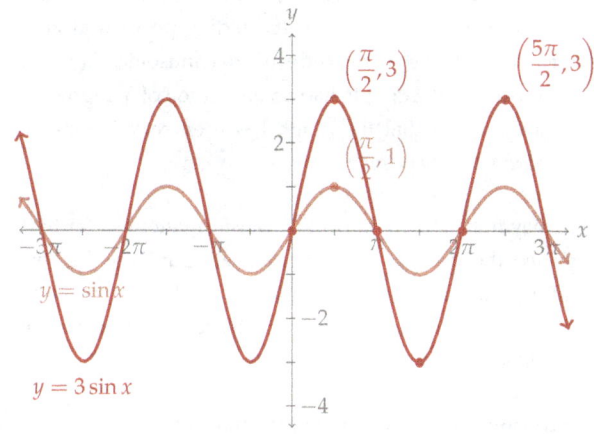

Since this transformation is affecting the graph in a vertical way, we may wish to rewrite the equation so that the modification is made on the y-variable:

$$\left(\frac{y}{3}\right) = \sin x.$$

This will also serve as a general rule; dividing the y-coordinate by a constant b will scale the graph vertically by a factor of b. Whether this makes the graph "taller" or "shorter" than the original will depend on the size of b.

Lesson 16

This type of transformation affects the maximum variation of the function's values from its starting position at zero, which we call the **amplitude** of the sinusoidal function. Our most recent example had an amplitude of 3, representing the amount that the graph has been scaled from the standard graph $y = \sin x$.

We may notice that none of the transformations so far have affected the **period** of our graphs; all graphs so far have had a period equal to 2π. However, scaling a sinusoidal curve in the horizontal direction will change this period, as we will soon see.

Example: Sketch the graph of the function
$$y = \sin\left(\frac{x}{2}\right).$$

We create a table of values:

$$\sin\left(\frac{1}{2}(0)\right) = 0 \qquad \sin\left(\frac{1}{2}\left(\frac{3\pi}{2}\right)\right) = \frac{\sqrt{2}}{2}$$

$$\sin\left(\frac{1}{2}\left(\frac{\pi}{2}\right)\right) = \frac{\sqrt{2}}{2} \qquad \sin\left(\frac{1}{2}(2\pi)\right) = 0$$

$$\sin\left(\frac{1}{2}(\pi)\right) = 1 \qquad \sin\left(\frac{1}{2}\left(\frac{5\pi}{2}\right)\right) = -\frac{\sqrt{2}}{2}$$

The points that we have just found are plotted below. The amplitude of this graph has not changed from the original sine curve; it still is bounded between -1 and 1.

However, the values that we have found do not even represent one complete period of this new function. Instead, it would take 4π radians before our values would start to repeat. So this modification to the original sine curve stretches our graph horizontally by a factor of 2, doubling the period.

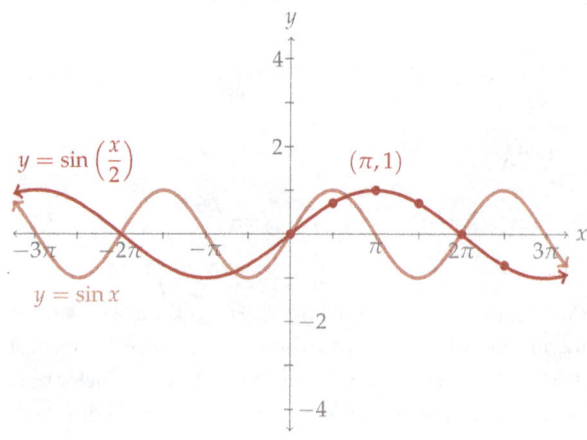

LESSON 16

Example: Sketch the graph of the function

$$y = \sin(2x).$$

Another table of values follows:

$$\sin(2(0)) = 0 \qquad \sin\left(2\left(\frac{3\pi}{2}\right)\right) = 0$$

$$\sin\left(2\left(\frac{\pi}{2}\right)\right) = 0 \qquad \sin(2(2\pi)) = 0$$

$$\sin(2(\pi)) = 0 \qquad \sin\left(2\left(\frac{5\pi}{2}\right)\right) = 0$$

It seems that all of these choices of angles give us a value of zero. But we certainly expect there to be some values that are non-zero, so perhaps we will plug in values in between the ones we have already evaluated:

$$\sin\left(2\left(\frac{\pi}{4}\right)\right) = 1 \qquad \sin\left(2\left(\frac{7\pi}{4}\right)\right) = -1$$

$$\sin\left(2\left(\frac{3\pi}{4}\right)\right) = -1 \qquad \sin\left(2\left(\frac{9\pi}{4}\right)\right) = 1$$

$$\sin\left(2\left(\frac{5\pi}{4}\right)\right) = 1 \qquad \sin\left(2\left(\frac{11\pi}{4}\right)\right) = -1$$

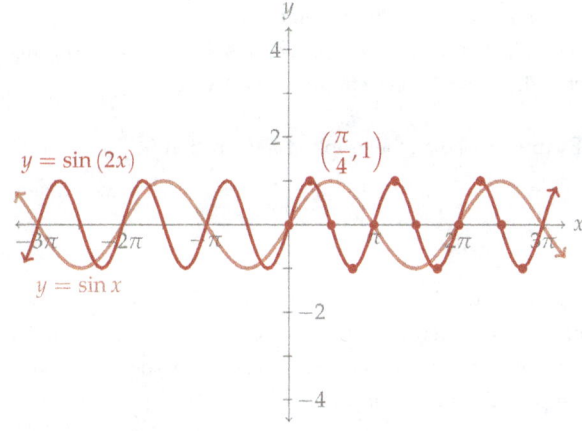

Above, we've plotted the twelve points that we've found so far, and filled in the graph for more points in between these. Note that the period of this function has been reduced by half, to equal π. We can summarize the effects of these scalings below.

- Replacing x with $\left(\frac{x}{a}\right)$ **scales** the graph **horizontally** by a factor of a.

- Replacing y with $\left(\frac{y}{b}\right)$ **scales** the graph **vertically** by a factor of b.

183

lesson 16

The last type of transformation is called a **reflection**. This transformation acts in the same way as a scaling, by multiplying either x or y by a common factor. However, in this case the factor is very specific: **negative one**.

Example: Sketch the graph of the function

$$y = -\sin(x).$$

Just as we did in the latest example, we'll again find values of y for some representative angles. To keep things simple, it is common to evaluate these functions at points that have simple values like 0 or ±1.

$$-\sin(0) = -(0) = 0 \qquad -\sin\left(\frac{3\pi}{2}\right) = -(-1) = 1$$

$$-\sin\left(\frac{\pi}{2}\right) = -(1) = -1 \qquad -\sin(2\pi) = -(0) = 0$$

$$-\sin(\pi) = -(0) = 0 \qquad -\sin\left(\frac{5\pi}{2}\right) = -(1) = -1$$

Of course, these are only some of the y-values that this function takes. But we can already see that for each point (x, y) on the standard graph of sine, the point $(x, -y)$ will be on the modified graph of the function $y = -\sin x$.

That is, all points on the original graph have been **reflected** vertically, across the x-axis, by negating their y-coordinates. We sometimes like to emphasize that this modification reflects the graph **vertically** by moving the negative to the other side of the equation with the vertical y-coordinate:

$$(-y) = \sin x.$$

The substitution of $(-y)$ for y will always have this effect.

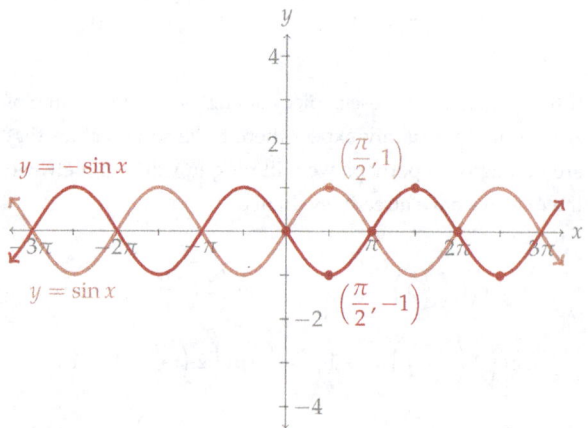

184

We might now suspect that substituting $(-x)$ for the horizontal variable x will have an analogous effect, reflecting the graph horizontally across the y-axis. This is indeed the case. However, for this particular example, we recall that sine is an **odd** function, and satisfies the identity

$$\sin(-x) = -\sin(x).$$

Therefore in this case, the horizontal and vertical reflections are one and the same.

For an example in which they are **not** the same, we can turn to the **even** function $y = \cos x$. For this function,

$$\cos(-x) = \cos(x).$$

In fact, we see in the graph at right that the horizontal reflection of the graph of cosine across the y-axis is exactly identical to the original. However, the vertical reflection across the x-axis **does** create a new graph, with equation

$$(-y) = \cos x.$$

Both reflections and scalings act by multiplying one of the variables by a scalar. However, reflections are usually considered an independent transformation.

To summarize our rules for reflections:

- Replacing y with $(-y)$ will **reflect** the graph vertically across the x-axis.

- Replacing x with $(-x)$ will **reflect** the graph horizontally across the y-axis.

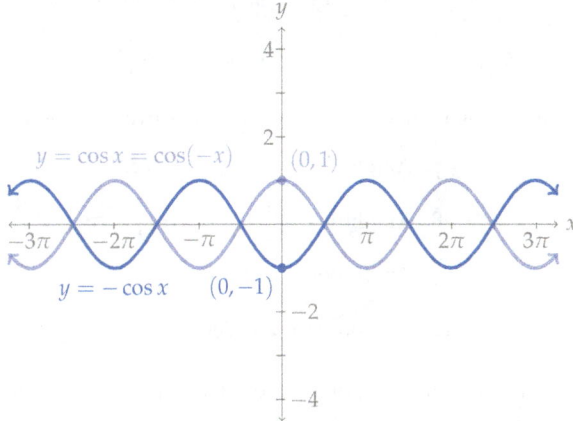

Now that we've seen the three types of transformations, we should point out that the **order** in which transformations are applied will usually affect the final outcome of the graph. For an example of this, see Exercise 23 of this lesson.

LESSON 16

EXERCISES

In Exercises 1–6, find both the amplitude and period of each sinusoidal curve.

1. $y = 3\sin(x)$
2. $y = \cos(4x)$
3. $y = -15\sin\left(\frac{x}{2}\right)$
4. $y = -2\sin(2\pi x)$
5. $y = \frac{3}{4}\sin\left(\frac{2}{3}x\right)$
6. $y = \pi\cos\left(\frac{\pi x}{3}\right)$

Write an equation for a sine curve with the given properties.

7. Amplitude = 2, Period = 3π
8. Amplitude = $\frac{1}{7}$, Period = $\frac{\pi}{2}$
9. Amplitude = 3, Period = 2
10. Amplitude = 2π, Period = $\frac{1}{3}$

Identify any graph transformations that have modified the standard graph of sine. Then sketch the given curve.

11. $y = \sin(2x)$
12. $y = \frac{5}{4}\sin(x)$
13. $y = -\sin\left(\frac{x}{4}\right)$
14. $y = 7\sin\left(x + \frac{\pi}{2}\right)$
15. $y = \sin(\pi x) + 3$
16. $y = 1 - \sin\left(\frac{\pi x}{6}\right)$

In Exercises 17–22, transform each function as indicated. It is not necessary to actually graph the functions.

17. $y = x^2$, shift graph right by 5 units.
18. $y = x^2$, shift graph down by 5 units.
19. $y = x^3 + 2x$, scale graph horizontally by a factor of 4.
20. $y = 4x^2 - 2x^3$, reflect graph across the y-axis.
21. $y = \cos(4x)$, scale graph vertically by a factor of $\frac{1}{2}$.
22. $y = \sin(x) + \cos(x)$, shift graph left by $\frac{\pi}{3}$ units.

23. Transform the graph of $y = \sin x$ by first shifting it horizontally to the left by $\pi/3$, then scaling it horizontally by a factor of 3. Then start over, first scaling horizontally by a factor of 3, and then shifting horizontally to the left by $\pi/3$. Sketch both graphs. Are they the same? Which operation corresponds to the expression $y = \sin(\omega(x - \phi))$? (The value ϕ in this expression is called the **phase shift**.)

24. Transform the graph of $y = \cos x$ by first shifting it up by 4 units, then reflecting it across the x-axis. Then start over, first reflecting across the x-axis, and then shifting it up by 4 units. Sketch both graphs. Are they the same?

LESSON 17

Quadratic Equations and Conic Sections

The goal of this lesson is to develop an algebraic tool that will we will use extensively for the remainder of this book. Historically, this tool is nearly as old as algebra itself. It has the power to completely solve any quadratic equation of one variable, even those that do not factor easily. It also allows us to simplify quadratic equations of two or more variables. This tool is called **completing the square**.

Before we demonstrate this method, we'll recall what it means for an equation to be quadratic. We've already factored a few of these equations in Lesson 8. In that lesson, we were solving equations that involved the trigonometric functions and had only one variable.

Definition: A **quadratic equation in one variable** is a polynomial equation whose individual terms have degree two or less. The **general form** of such an equation is

$$ax^2 + bx + c = 0.$$

Here x is an unknown variable, and the real numbers a, b, and c are fixed constants called **coefficients**. We sometimes refer to a as the **quadratic** coefficient, to b as the **linear** coefficient, and to c as the **constant** coefficient.

We often call an equation quadratic even if the variable x is a function in its own right, as was the case in Lesson 8.

Lesson 17

Among the quadratic equations we've just defined are the **perfect squares**: the equations of the form $(x+h)^2 = 0$. These form a small subset of all the quadratic equations, and we'd like to be able to identify when a given equation is or is not a square.

We'll quickly remind ourselves that $(x+h)^2$ is **not** equal to $x^2 + h^2$, since we must distribute:

$$(x+h)^2 = (x+h)(x+h)$$
$$= x(x+h) + h(x+h)$$
$$= x^2 + xh + hx + h^2$$
$$= x^2 + 2hx + h^2.$$

However, from this calculation we can see an important fact. Note the relationship between the linear and constant coefficients: If a quadratic equation $ax^2 + bx + c = 0$ is a perfect square, and we set $c = h^2$, then we must have that $b = 2h$. That is, b and c should be related by

$$c = \left(\frac{b}{2}\right)^2.$$

This fact furnishes us with the tool we mentioned earlier. We'll demonstrate how to use it by solving a quadratic equation of one variable.

Example: Find the values of x that satisfy the equation

$$x^2 - 2x - 15 = 0.$$

We immediately see that the left-hand side of this equation is not a perfect square, since $c = -15$, and

$$-15 \neq \left(\frac{-2}{2}\right)^2.$$

However, we can "complete" the left-hand side by adding on an appropriate constant. In preparation for this, we move the existing constant to the right-hand side:

$$x^2 - 2x = 15.$$

Next, since $b = -2$, we calculate the quantity

$$\left(\frac{b}{2}\right)^2 = \left(\frac{-2}{2}\right)^2 = (-1)^2 = 1.$$

According to our earlier calculation, here $h = -1$ and

$$h^2 = (-1)^2 = 1.$$

We will add this to both sides of our equation:

$$x^2 - 2x + 1 = 15 + 1.$$

LESSON 17

Now, the left-hand side **is** a perfect square; specifically, it is equal to $(x-1)^2$. You can (and should) multiply out this product using the distributive property to verify that they are equal. We can also simplify the constants on the right-hand side to obtain

$$(x-1)^2 = 16.$$

This equation is now in a position to be solved, by taking the square root of both sides. Recall that when doing so, we must account for both positive and negative square roots.

$$(x-1)^2 = 16$$
$$x - 1 = \pm\sqrt{16}$$
$$x = 1 \pm 4$$
$$x = 5 \quad \text{or} \quad x = -3.$$

Plugging these values into our original equation will verify that they really are solutions to the original equation.

Again, this tool is extremely useful in this context because it can be used for **any** single-variable quadratic equation whatsoever, much like the famous quadratic formula. But unlike the formula, completing the square can be used in several contexts other than solving quadratic equations.

Steps for Completing the Square:

1. Gather the like terms of the unknown variable and identify the linear coefficient.

2. Divide the linear coefficient by two, then square it to find the appropriate constant h^2.

3. Add and subtract this constant from the equation.

4. Simplify.

In the remainder of this text, we will be applying this tool to equations involving more than one variable. Recall that such equations can generally not be solved for any particular solution, but instead define a **set** of solutions that can be visualized by a graph in a coordinate system.

Definition: A **quadratic equation in two variables** is a polynomial equation whose individual terms have degree two or less. The **general form** of such an equation is

$$ax^2 + by^2 + cx + dy + exy + f = 0.$$

To simplify matters in this course, we will generally take the coefficient e to be zero, so that we have no "mixed" term.

Lesson 17

Example: Complete the square for both variables:

$$x^2 + y^2 - 6x + 10y - 4 = 0.$$

Here we will first separate the x-terms and y-terms, and also move the constant term to the right-hand side:

$$x^2 - 6x + y^2 + 10y = 4.$$

Next, we see that the two linear coefficients for x and y are $b = -6$ and $b = 10$, respectively. Dividing each in half and squaring, we obtain the two needed constant terms. The two quadratic expressions on the left-hand side are

$$x^2 - 6x + \left(\frac{-6}{2}\right)^2,$$

$$y^2 + 10y + \left(\frac{10}{2}\right)^2.$$

We can factor these into two completed squares, and add the appropriate constants to the right-hand side as well.

$$x^2 - 6x + (-3)^2 + y^2 + 10y + (5)^2 = 4 + (-3)^2 + (5)^2$$

$$(x-3)^2 + (y+5)^2 = 4 + 9 + 25$$

$$(x-3)^2 + (y+5)^2 = 38.$$

It may not be apparent how this form of the equation is any better or worse than the original form in which it was given. However, this is the form of a quadratic equation that we will want to obtain in the future.

The next example differs from the previous one, as only one of the variables has a quadratic term. The other variable has only a linear, degree-one term.

Example: Complete the square for x:

$$x^2 - 8x + 5y + 3 = 0.$$

We first add the constant term necessary to complete the square for x. To do this, identify the linear coefficient of the variable x, divide it by two, and square:

$$x^2 - 8x + \left(\frac{-8}{2}\right)^2 + 5y + 3 = \left(\frac{-8}{2}\right)^2.$$

We'll simplify this expression by grouping the like terms:

$$x^2 - 8x + (-4)^2 + 5y + 3 = (-4)^2$$

$$(x-4)^2 + 5y + 3 = 16$$

$$(x-4)^2 + 5y = 13.$$

Finally, we'll look at an example in which the quadratic coefficients are not equal to one.

Example: Complete the square for both variables:

$$5x^2 - 16y - 27 = 2y^2 - 15x.$$

Though this expression is written a bit haphazardly, with terms on both sides of the equation, each variable still has both a linear and a quadratic term. We will group them together before completing each square. It does not matter which side keeps the constant term.

$$5x^2 + 15x - 27 = 2y^2 + 16y.$$

Now, we isolate the quadratic coefficients by factoring them out of each side. Notice that we continue to leave the constant term alone during this operation.

$$5(x^2 + 3x) - 27 = 2(y^2 + 8y).$$

We are now in a position to complete the square, within both sets of parentheses. When we do, we will need to add the appropriate constants to both sides of the equation. It is critical to remember that anything added inside a set of parentheses is also being multiplied by the constant in front.

We account for this when adding the corresponding constant to the opposite side of the equation. Inside the parentheses, we've added a constant, but outside we've added the product of the constant and the quadratic coefficient. The original constant has not been modified.

$$5\left(x^2 + 3x + \frac{9}{4}\right) + 2(16) - 27 = 2(y^2 + 8y + 16) + 5\left(\frac{9}{4}\right).$$

The quadratic expressions in each set of parentheses are perfect squares, and the remaining constants can be grouped together and simplified.

$$5\left(x + \frac{3}{2}\right)^2 + 32 - 27 = 2(y + 4)^2 + \frac{45}{4}$$

$$5\left(x + \frac{3}{2}\right)^2 - 2(y + 4)^2 = \frac{45}{4} - 5$$

$$5\left(x + \frac{3}{2}\right)^2 - 2(y + 4)^2 = \frac{25}{4}.$$

Again, it does not seem as though this is a significant improvement over the original form of this equation. But when equations are written in this form, we can more easily see the graph transformations that might help us identify the solution set of this equation in a coordinate system.

LESSON 17

In the next few lessons we will be using quadratic equations and completing the square to conduct a detailed study of three important geometric curves. These curves can occur in many natural ways, but the most common method of obtaining their shapes is by intersecting a flat plane with a right circular cone.

The type of curve generated by this intersection is determined by the angle of the plane relative to the cone that it intersects. It is because of this description that the curves are collectively referred to as **conic sections**. Individually, the three types of curves are referred to as **ellipses**, **parabolas**, and **hyperbolas**.

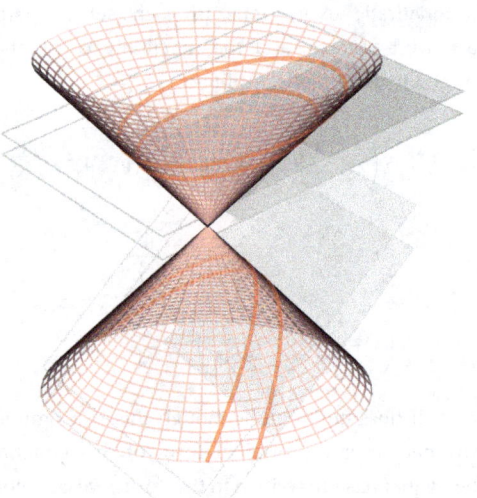

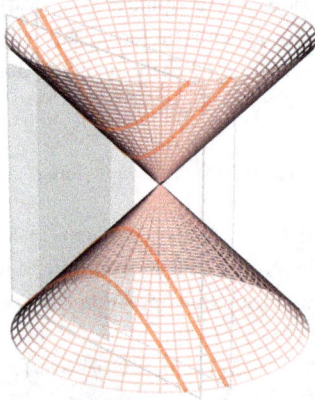

This description shows the most common way that conic sections arise in nature. But without a coordinate system in place, or some sort of accurate description of the cone in question, we would find it quite difficult to measure dis-

LESSON 17

tances between points or other geometric properties of conic sections. For this reason, some alternative definitions are often used.

One such set of definitions was given to us by Apollonius of Perga, who lived in ancient Greece from around 262 to 190 BCE. It was Apollonius who gave these curves their names. Though he knew them to be generated by intersections with a cone, for formal purposes he defined each of them with the following descriptions.

Definition: Let L be a line and F be a point not on the line. A **parabola** is the set of points that are equidistant from F and L. The point F is called the **focus** of the parabola, and the line L is called the **directrix**.

Definition: Let P and Q be two points. An **ellipse** is the set of points for which the sum of the distances to both P and Q is some fixed constant. The points P and Q are called the **foci** of the ellipse.

Definition: Let P and Q be two points. A **hyperbola** is the set of points for which the difference of the distances to P and Q is some fixed constant. The points P and Q are called the **foci** of the hyperbola.

While each of these definitions is still completely geometric in nature, they at least offer a bit more information about how the points on each curve are related to each other, and they highlight important features of each curve that we did not see in our previous description. We might also notice that each of these new definitions is entirely independent of the others. Though they seem to be similar in their wording, this fact alone might not suggest that they are as closely related as we know them to be.

It is quite remarkable that Apollonius was able to logically connect these various definitions. After all, he was without most of the modern mathematical tools that we currently have available.

There were two key developments that allowed the description of conics to progress past Apollonius. The first was the development of algebra and algebraic equations, which began in India with Aryabhata (476–550 CE), and was more fully developed in the Abbasid Caliphate by Muḥammad ibn Mūsā al-Khwārizmī (c. 780–850 CE).

The methods of algebra introduced the idea of assigning a letter to an unknown quantity, the symbolism to create equations that represent the relationships between the let-

Lesson 17

ters, and the rules by which the resulting equations could be manipulated to identify the unknown.

The second development was the introduction of coordinate systems, and the notion of viewing the solutions of an algebraic equation as geometric objects in a coordinate system. As we saw in Lesson 12, this work was mainly due to René Descartes (1596–1650 CE), occurring nearly two thousand years after the death of Apollonius.

These new tools of mathematics give us a third, more modern way to describe conic sections, as the graphs of algebraic equations. Specifically, the conic sections are those graphs that arise from **quadratic** equations.

Theorem: Suppose that we have a quadratic equation:

$$ax^2 + by^2 + cx + dy + f = 0.$$

If either of the quadratic coefficients a or b is zero, leaving an equation with only one quadratic term, then the graph of the equation will form a **parabola** in the x,y-plane.

If both of the quadratic coefficients are nonzero and have the same sign, then the graph of the equation will form an **ellipse** in the x,y-plane.

If both of the quadratic coefficients are nonzero and have opposite signs, then the graph of the equation will form a **hyperbola** in the x,y-plane.

It's good to note that these descriptions may include some defunct cases. For example, consider the equation

$$x^2 + y^2 + 4 = 0.$$

The above descriptions state that graphing the solutions to this equation in the x,y-plane should leave us with an ellipse. However, by rewriting the equation

$$x^2 + y^2 = -4,$$

we see that the left-hand side is always positive, and the right-hand side is negative. Therefore, in this case the equation has no solutions to graph.

In Lessons 18–20 we'll study each type of conic section individually. For each case, we'll start with a set of points satisfying the ancient geometric definitions, place these points into a coordinate system, and then use algebra to obtain the type of equations we have just described. Along the way we'll also identify the essential geometric features of each curve with complete exactness.

LESSON 17

EXERCISES

For Exercises 1–3, identify the quadratic, linear, and constant coefficients of each quadratic equation.

1. $x^2 + 2x + 5 = 0$ **2.** $-3x^2 + 3 = 0$ **3.** $5x^2 - \pi x = 0$

Decide whether or not each expression is a perfect square.

4. $x^2 + 4x + 4$ **5.** $a^2 + 14a + 28$ **6.** $z^2 - 6z - 9$

7. $k^2 - k + \frac{1}{4}$ **8.** $m^2 - 10m + 25$ **9.** $b^2 - 3b + \frac{9}{2}$

For Exercises 10–15, find the value of c that makes each quadratic expression into a perfect square.

10. $x^2 + 6x + c$ **11.** $h^2 + 9h + c$ **12.** $r^2 - \frac{1}{2}r + c$

13. $n^2 + 50n + c$ **14.** $m^2 - \frac{4}{5}m + c$ **15.** $a^2 + 11a + c$

Solve each equation by completing the square.

16. $x^2 + 4x = 96$ **17.** $t^2 + 8t - 20 = 0$

18. $c^2 + 3c - 180 = 0$ **19.** $4y^2 + 19y - 5 = 0$

20. $m^2 + \frac{7}{3}m + \frac{2}{3} = 0$ **21.** $2a^2 - 3a + 4 = 0$

In Exercises 22–26, complete the square for any quadratic variables in each equation.

22. $x^2 + y^2 - 2x - 4y + 5 = 0$

23. $x^2 - 6x + 2y + 9 = 0$

24. $9x^2 - y^2 - 72x + 8y + 119 = 0$

25. $x^2 - 9y^2 + 10x + 18y + 7 = 0$

26. $x^2 - y^2 + z^2 + 3x + 8y - 7z = 0$

State whether the graph of each equation is a parabola, an ellipse, or a hyperbola.

27. $9x^2 - 25y^2 + 50y = 18x$ **28.** $3x^2 + 4y^2 + 8y = 8$

29. $x^2 - 6x - y + 34 = 0$ **30.** $y^2 - 2x^2 - 16 = 0$

31. $(x-3)^2 + (y-4)^2 = (x+1)^2$ **32.** $4x^2 + 9y^2 = 16$

33. Without knowing the values of the constants a, b, or c, by completing the square, find an expression for the solutions of the following equation. Show all of your work.

$$ax^2 + bx + c = 0.$$

LESSON 18

Parabolas

In Lesson 17 we were left with three competing definitions of a parabola. The first definition was not terribly useful; without knowing anything specific about the cone or the plane that intersects it, we are not left with much to go on.

The second definition, due to Apollonius, at least detailed some information about the distances between the focus, directrix, and points on a parabola. This tells us something about the shape of a parabola relative to these features, but we still have no way of determining the position or orientation of this geometric object.

Our third definition is much better. By defining the parabola as the graph of a particular form of equation, we naturally introduce a coordinate system in which to measure our object.

However, while the shapes produced by these geometric and algebraic definitions may **seem** similar, we still may not be convinced that they actually **are** the same shape. For this we will need to do a bit more work, placing Apollonius' parabola into Descartes' coordinate system, and using the algebra of al-Khwārizmī to connect the definitions.

LESSON 18

Recall Apollonius' geometric definition of a parabola:

Definition: Let L be a line and F be a point not on the line. A **parabola** is the set of points that are equidistant from F and L. The point F is called the **focus** of the parabola, and the line L is called the **directrix**.

Our immediate goal is to show that, given a set of points in the x,y-plane that satisfies Apollonius' definition, the equation describing this set will be a quadratic equation of the type we have discussed. Along the way, we hope to learn something about the position of the focus and directrix in relation to the rest of the parabola.

Let's start with a focus at the point $F = (h, c)$, shown in the image at right. There is also a directrix L with equation $y = -c$. We will call S the set of points that lie equidistant from F and L.

We notice a few things from the image. First, the minimum distance from a point P is the length of a segment that is perpendicular to L, so the endpoints of the segment have the same x-coordinate. Second, we have chosen the y-coordinates of F and L so that the point $(h, 0)$ lies directly in between the two, and is therefore a point in S.

This was done to simplify the following calculation, but it was not a necessary choice. The distance between P and the point $(x, -c)$ on L can be written

$$\mathrm{dis}(P, L) = \sqrt{(x-x)^2 + (y-(-c))^2}.$$

Likewise, the distance between P and F can be written

$$\mathrm{dis}(P, F) = \sqrt{(x-h)^2 + (y-c)^2}.$$

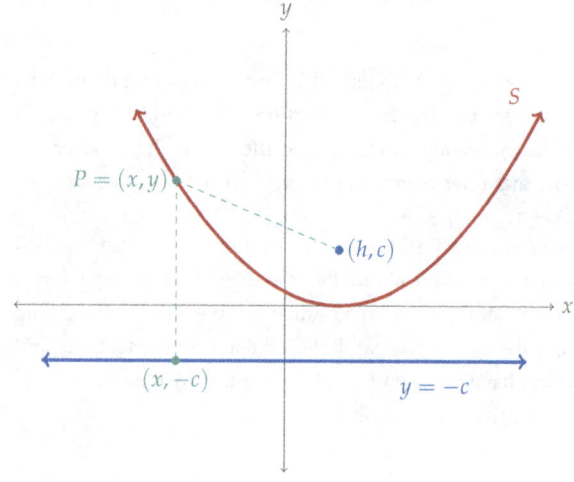

The point P lies in S, so these distances are assumed to be equal. Therefore, we set these two expressions equal to each other and simplify:

$$\text{dis}(P, L) = \text{dis}(P, F)$$
$$\sqrt{(x-x)^2 + (y+c)^2} = \sqrt{(x-h)^2 + (y-c)^2}$$
$$(0)^2 + (y+c)^2 = (x-h)^2 + (y-c)^2$$
$$y^2 + 2cy + c^2 = (x-h)^2 + y^2 - 2cy + c^2$$
$$2cy = (x-h)^2 - 2cy$$
$$4cy = (x-h)^2.$$

As expected, this equation is quadratic in only one of its variables, so it represents a parabola in the algebraic sense.

From our image we can see that the minimum possible distance from a point on the parabola to the vertex is precisely c. Therefore the role of the number 4 in the last equation above is significant. Also significant is the single point that lies this distance away from both the focus and directrix. For any other distance $d > c$, the points lying distance d from the focus occur in pairs. The importance of this unique point is indicated in the following definition.

Definition: The point of the parabola having the minimum distance to the directrix (and also to the focus) is called the **vertex**. The line passing through both the vertex and the focus of the parabola is called the **axis of symmetry**.

In other words, the vertex lies a distance of c away from both the focus and the directrix, along the axis of symmetry. We will usually try to put these equations into the form shown, separating off the factor of 4, to identify c.

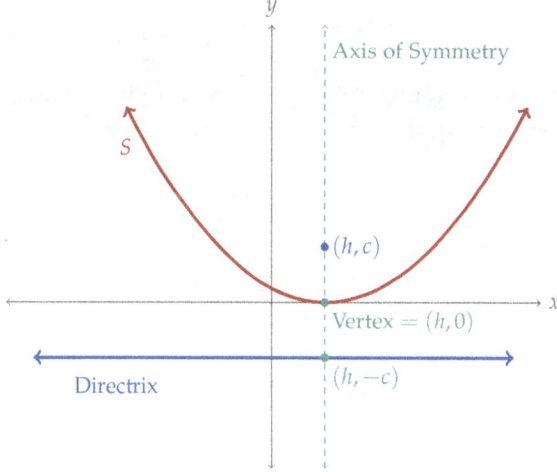

Lesson 18

Example: Find the focus, directrix, vertex, and axis of symmetry of the parabola given by

$$4x^2 + 9 = 48y - 12x.$$

As a very first step, we will separate the variables x and y, placing them on opposing sides of the equation. To simplify our work in completing the square later, we can also divide the quadratic coefficient to the side with y:

$$48y = 4x^2 + 12x + 9$$

$$12y = x^2 + 3x + \frac{9}{4}.$$

Now we see that the right-hand side is actually a complete square already, and nothing more needs to be added:

$$12y = \left(x + \frac{3}{2}\right)^2.$$

Finally, we factor a 4 from the coefficient of y. This identifies the value of c:

$$4(3)y = \left(x + \frac{3}{2}\right)^2.$$

Thus the vertex of our parabola is at $(h, 0) = \left(-\frac{3}{2}, 0\right)$.

This parabola will open upward, and our axis of symmetry will be the vertical line passing through the vertex. That is, the axis of symmetry has the equation

$$x = -\frac{3}{2}.$$

The focus lies on this axis of symmetry, $c = 3$ units above the vertex, at

$$(h, c) = \left(-\frac{3}{2}, 3\right).$$

Finally, the directrix is the horizontal line lying $c = 3$ units below the vertex, having equation $y = -3$.

We leave this example to give another definition relevant to parabolas. This feature determines how **wide** a parabola opens. In cases where the parabola crosses an axis, we might be able to use the intercepts for this purpose; knowing the location of the vertex and any intercepts should be enough to graph a parabola exactly. However, this solution may not work in all situations.

Definition: The **focal chord** (or **latus rectum**) of a parabola is the line segment that passes through the parabola's focus, is parallel to the parabola's directrix, and has its endpoints on the parabola itself.

To find the length of this focal chord and the locations of its endpoints, let's return to our previous standard form:

$$4cy = (x-h)^2.$$

The endpoints of the focal chord lie on both the chord itself and on the parabola. Intersecting the horizontal line passing through the focus, $y = c$, with the parabola itself, we find their x-coordinates:

$$4cy = (x-h)^2$$
$$4c(c) = (x-h)^2$$
$$4c^2 = (x-h)^2$$
$$\pm 2c = x - h$$
$$h \pm 2c = x.$$

Recall that $y = c$ in this case, so the two endpoints of the focal chord will be $(h-2c, c)$ and $(h+2c, c)$.

At right we have an image depicting all of the main features of a parabola, and their positions relative to each other. Note that this image does not indicate either a position or an orientation in any coordinate system. All parabolas will have these same features, in the same relative positions.

The proportions of a parabola and its focal chord will always depend exclusively on the value of c, while the position and orientation of a parabola in space are defined by other factors.

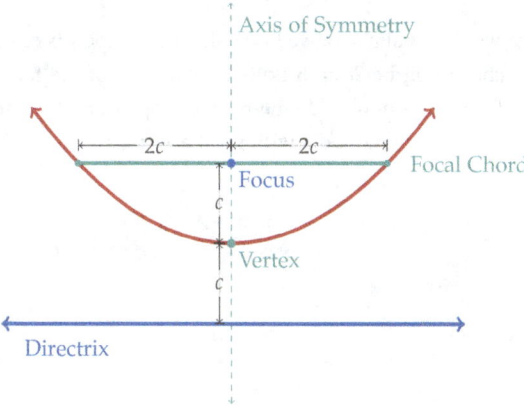

We will explore those other factors now, with a number of examples having to do with the equations and graphs of parabolas. In particular, we'll be paying attention to how our standard graph transformations affect the position, orientation, and shape of the parabolas.

LESSON 18

Example: Earlier, we found the focus, directrix, vertex, and axis of symmetry of the parabola $4x^2 + 9 = 48y - 12x$. The standard form for this parabola was

$$4(3)y = \left(x + \frac{3}{2}\right)^2.$$

Since we find that $c = 3$, we know that the endpoints of the focal chord will be $2c = 6$ units to each side of the focus. This determines how wide the parabola opens, and we are able to graph the parabola with great accuracy.

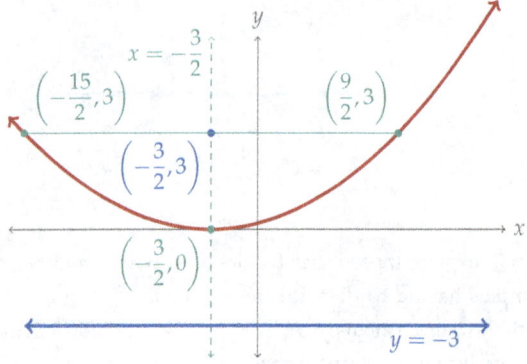

Furthermore, the sign of the linear y-coefficient determines the direction that the parabola opens. If we change the sign, this will **reflect** the parabola across the horizontal axis passing through the vertex.

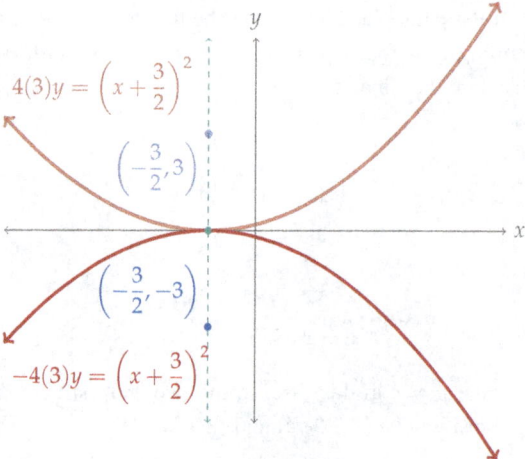

Note that this does not change the vertical axis of symmetry, nor does it change the vertex of the parabola. This change does reflect all other features of the parabola, however.

Another way that we can transform this graph is by **shifting** it vertically. Recall that this involves making a change to the y-variable in the equation.

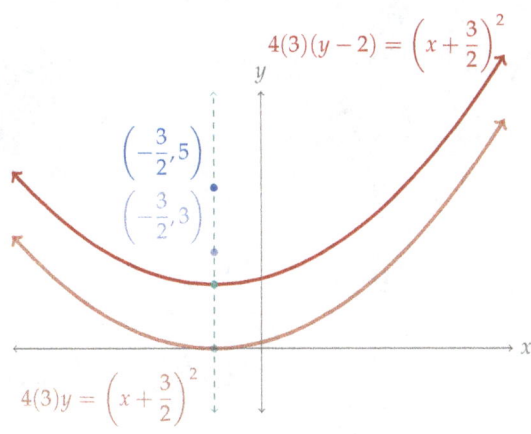

We might also choose to **scale** this equation in either the horizontal or vertical direction. As an example, we might stretch the graph of our original parabola vertically by a factor of 2. We do this algebraically by replacing the variable y with the expression $y/2$.

Performing this replacement and simplifying, we find that this new constant is absorbed by the value c, moving our focus and altering the length of our focal chord:

$$4(3)\left(\frac{y}{2}\right) = \left(x + \frac{3}{2}\right)^2$$

$$4\left(\frac{3}{2}\right)(y) = \left(x + \frac{3}{2}\right)^2.$$

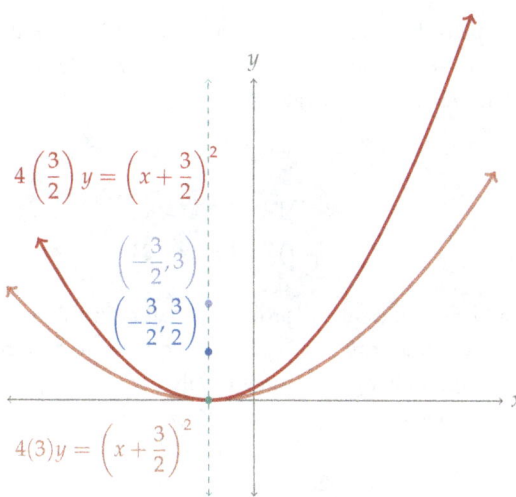

LESSON 18

Example: Find the equation of the parabola having its vertex at $(-3, 1)$ and its focus at $(2, 1)$.

Here, the vertex and focus lie on the same horizontal line $y = 1$, which is therefore the axis of symmetry for this parabola. The parabola will open **horizontally**, and with the focus to the right of the vertex, it must open in the direction of the positive x-axis.

The distance between the vertex and focus is $c = 5$, and with the vertex being $(h, k) = (-3, 1)$, we can build up the standard form for this parabola:

$$(y - k)^2 = 4c(x - h)$$
$$(y - 1)^2 = 4(5)(x - (-3))$$
$$(y - 1)^2 = 4(5)(x + 3).$$

It would be sufficient to leave the equation in this form. But if desired, we could also place the equation in **general form** by multiplying everything together and moving it to one side of the equation:

$$y^2 - 2y + 1 = 20x + 60$$
$$y^2 - 2y - 20x - 59 = 0.$$

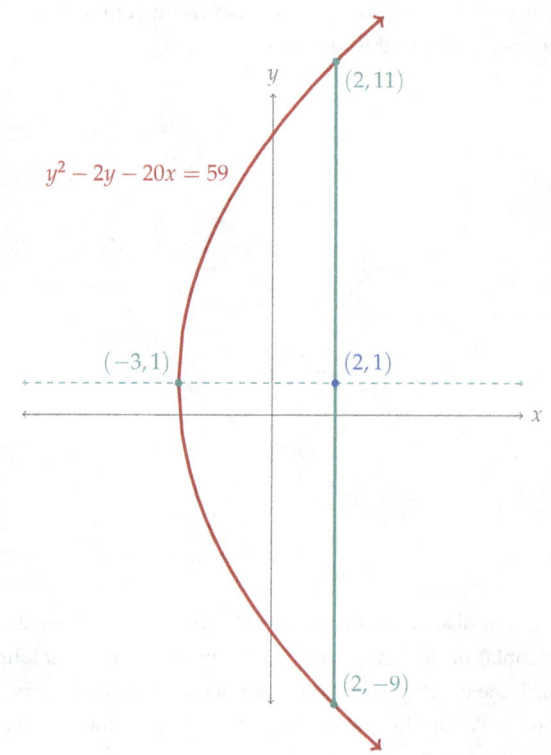

204

LESSON 18

Now that we have seen both horizontal and vertical parabolas, we'll collect each of the standard forms, along with the locations of their defining features, in one location.

Standard Form	$\pm 4c(y-k) = (x-h)^2$
Vertex	(h,k)
Axis of Symmetry	$x = h$
Focus	$(h, k \pm c)$
Directrix	$y = k \mp c$
Focal Chord	From $(h-2c, k \pm c)$ to $(h+2c, k \pm c)$

Standard Form	$(y-k)^2 = \pm 4c(x-h)$
Vertex	(h,k)
Axis of Symmetry	$y = k$
Focus	$(h \pm c, k)$
Directrix	$x = h \mp c$
Focal Chord	From $(h \pm c, k-2c)$ to $(h \pm c, k+2c)$

Any quadratic equation that has only one quadratic term may be placed into one of these standard forms by completing the square, and from there we can locate its features. We demonstrate this process with another example.

Example: Find the vertex, focus, and directrix:

$$y^2 - 8y - 8x - 8 = 0.$$

We immediately notice that the variable y has a quadratic term while x does not, so this parabola will open horizontally. We place the equation into standard form by completing the square.

$$y^2 - 8y = 8x + 8$$
$$y^2 - 8y + 16 = 8x + 8 + 16$$
$$(y-4)^2 = 8x + 24$$
$$(y-4)^2 = 8(x+3)$$
$$(y-4)^2 = 4(2)(x+3).$$

We can now see that the vertex is at $(-3, 4)$, and the parabola opens horizontally in the positive x-direction. The focus lies 2 units from the vertex in this direction, at $(-1, 4)$. The directrix lies 2 units in the opposite direction, being the vertical line $x = -5$.

LESSON 18

EXERCISES

For Exercises 1–4, write each equation in standard form:
$$\pm 4c(y-k) = (x-h)^2.$$

1. $x^2 + 30 = 10y$
2. $y - 33 = x^2 - 6x$
3. $y + 24x = 3x^2 + 50$
4. $y = \frac{1}{2}x^2 - 3x + \frac{19}{2}$

For Exercises 5–8, write each equation in standard form:
$$(y-k)^2 = \pm 4c(x-h).$$

5. $x = y^2 + 8y + 20$
6. $x = y^2 - 14y + 25$
7. $x = 5y^2 - 25y + 60$
8. $x = \frac{1}{4}y^2 - \frac{1}{2}y - 3$

The focus and directrix of several parabolas are given below. Write the equation of each parabola and sketch its graph.

9. $(8,0)$, $y = 4$.
10. $(6,2)$, $x = 4$.
11. $(3,-1)$, $x = -2$.
12. $(0,3)$, $y = -1$.
13. $(2,5)$, $y = 3$.
14. $(3,5)$, $x = -7$.

For Exercises 15–20, find the vertex, axis of symmetry, focus, directrix, and the direction that the parabola opens. Then find the length of the focal chord and sketch the graph.

15. $4y = x^2$
16. $y = x^2 - 6x + 11$
17. $(y-8)^2 = -4(x-4)$
18. $(x-8)^2 = \frac{1}{2}(y+1)$
19. $y^2 = -8x$
20. $(y-3)^2 = 4x - 4$

21. Write the equation of the parabola that has its vertex at $(5,-1)$, and its focus at $(3,-1)$. Then sketch its graph.

22. A parabola has its vertex at $(-7,4)$, and the endpoints of its focal chord lie at $(-1,1)$ and $(-13,1)$. Find the equation of the parabola and sketch its graph.

23. Suppose that a parabola is skew to the coordinate axes, so that it has its vertex at the origin and its focus at $(1,1)$. Find the equation of the directrix of this parabola.

24. Using Apollonius' definition of a parabola, and the focus and directrix from Exercise 23, find the equation of that parabola.

206

LESSON 19

Ellipses

Of the conic sections, the **ellipse** is likely to be the most familiar to us. After all, **circles** have been a central part of our study for the bulk of this text. We've even defined the six trigonometric functions themselves as the ratios of the coordinates of points on a circle. The Unit Circle later played a central role in our understanding of the arc-functions. Even later, concentric circles became the basis for the polar and cylindrical coordinate systems.

Of course, an ellipse is a more general object than a circle. If the connection between them is still elusive, we might look again at Apollonius' definition of the ellipse.

Definition: Let P and Q be two points. An **ellipse** is the set of points for which the sum of the distances to both P and Q is some fixed constant. The points P and Q are called the **foci** of the ellipse.

If we consider a circle to be the set of points a fixed distance from its center, then we might think of an ellipse as a "circle with two radii." Or conversely, we might consider a circle to be a special type of ellipse, in which the two foci lie at the same point. In this case the common focus is called the **center**, and the fixed constant referred to in the above definition is twice the circle's radius.

LESSON 19

We might also recall the third geometric definition of an ellipse, which describes it as the intersection of a right circular cone with a flat plane. Even in this description, we can still connect circles and ellipses. Here, a circle is the intersection of the cone with a plane that is completely perpendicular to the cone's axis of symmetry, as shown in the lower half-cone of the image below. Of course, the intersection shown in the upper half-cone does not form a circle, but rather an ellipse.

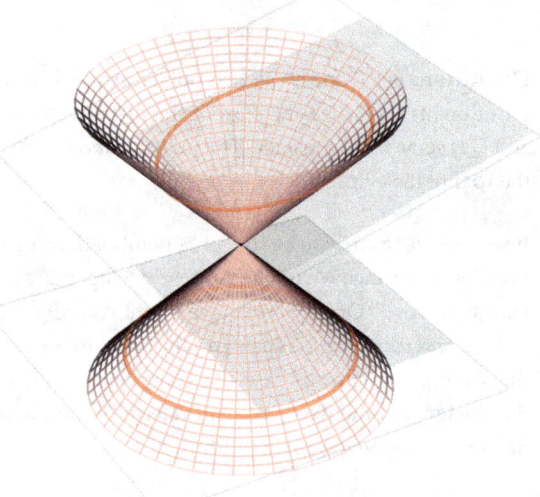

From all of these considerations it would certainly seem plausible that an ellipse would be nothing more than a transformed circle. Let's explore this idea by transforming a circle in a few different ways.

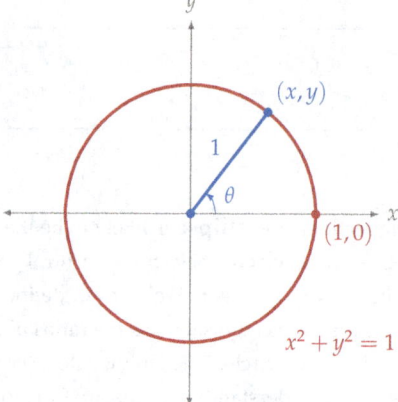

We remember from several of our previous lessons that a Unit Circle of radius $r = 1$, centered at the origin, will have equation $x^2 + y^2 = 1^2$. With this as a starting point, we might consider applying some transformations to this graph, changing its form.

For example, we can **scale** the graph of the circle, perhaps stretching it horizontally by a factor of two. We perform this action by replacing the variable x with the expression $x/2$, as is shown in red in the figure.

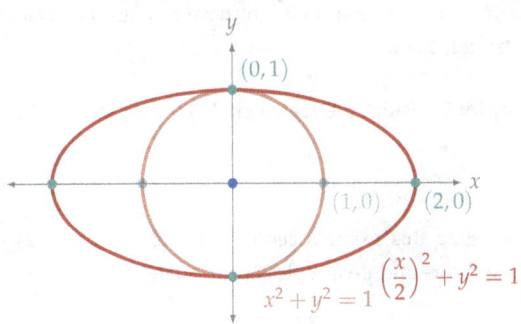

Since this graph is centered at the origin, this scaling has not moved the center at all. This is an important consideration when transforming graphs, and the reason that we typically perform any necessary translations as a final step.

We can stretch the graph vertically as well, which we do by replacing the variable y with the expression $y/3$. This is shown in the image at right, which has been transformed both horizontally and vertically.

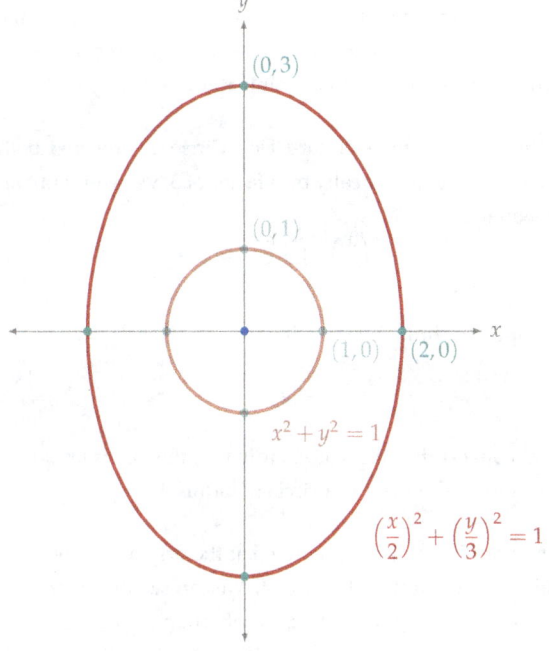

You will notice how these transformations move the four intercepts of the graph. The distance from the center to these intercepts can be easily seen in the denominators of the two squared terms of each equation.

Lesson 19

In the case that both of these denominators are equal, so that the graph is scaled horizontally and vertically by the same amount, we will be able to multiply the equation through by this common denominator to simplify it.

For example, if the standard Unit Circle is stretched both horizontally and vertically by a factor of 3, we would obtain the equation
$$\left(\frac{x}{3}\right)^2 + \left(\frac{y}{3}\right)^2 = 1.$$

Multiplying through by 9 will give us
$$x^2 + y^2 = 9.$$

If we convert this to polar coordinates, this becomes $r = 3$, which indeed describes a circle of radius 3.

For circles like this one, centered at the origin, a **reflection** will have no effect on the graph. We can see this geometrically but also algebraically, as a reflection about the x-axis would involve making the y-variable negative. But this will not alter the equation in any way:
$$x^2 + (-y)^2 = 9$$
$$x^2 + y^2 = 9.$$

However, suppose that a circle has its center away from the origin. Because a circle is symmetric both horizontally and vertically, a reflection will not change the shape or size of the circle. But it will reflect this center, which effectively **shifts** the ellipse to a different location in space. Therefore any effect that a reflection might have can also be achieved by a **translation**.

Example: Consider the scaled circle that we saw earlier:
$$\left(\frac{x}{2}\right)^2 + \left(\frac{y}{3}\right)^2 = 1.$$

We translate this graph's center from $(0,0)$ to $(1,-2)$ by shifting the x- and y-variables appropriately:
$$\left(\frac{x-1}{2}\right)^2 + \left(\frac{y+2}{3}\right)^2 = 1.$$

The graph of this ellipse is shown in the image at right. It has the same size and shape as the previous ellipse with its center at the origin. If we wished, we could either reflect this graph across the x-axis, or we could shift it vertically by 4 units, and both should have the same effect. Indeed, it is simple to verify that the following expressions are equal:
$$\left(\frac{x-1}{2}\right)^2 + \left(\frac{(-y)+2}{3}\right)^2 = \left(\frac{x-1}{2}\right)^2 + \left(\frac{y-2}{3}\right)^2.$$

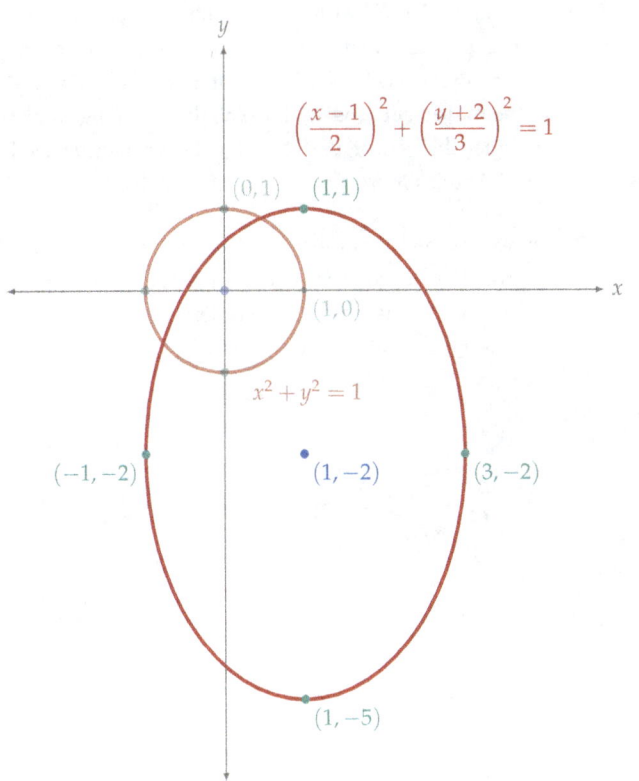

Note how this translation affected the four points that used to be the intercepts of the graph with the x- and y-axes. These points retain their distance from the center in their respective directions, so we are still able to identify them. These four points are called the **vertices** of the graph. Specifically, the two points that lie furthest from the center are called the **major vertices**, and the two lying closest to the center are called the **minor vertices**.

The two major vertices in this case were formerly the y-intercepts of the graph. These are still 3 units away from the center, as indicated by the denominator in the y-term of the equation. Therefore we can identify them as $(1, -2 \pm 3)$, or $(1, 1)$ and $(1, -5)$. Similarly, the two minor vertices lie at $(-1, -2)$ and $(3, -2)$.

Now that we've transformed a circle into various shapes and identified some of its features, we should approach the question of whether a "transformed circle" is actually an ellipse, according to the definition of Apollonius. These modifications certainly **appear** to give us the right shape. But we still have no way of knowing where the foci of such a transformed circle might lie. To make the connection, we'll do the same sort of calculation that we did for parabolas in Lesson 18.

LESSON 19

Starting from scratch, we'll place the two foci in a coordinate system, write expressions for the distances from each focus to a point on the ellipse, and set up an equation to simplify. This calculation is a bit longer than the one for parabolas, but it involves a certain strategy for solving equations with multiple square roots that is good to see.

For simplicity, we will establish our coordinate system with its origin at the center of the ellipse, and place the foci c units away in a horizontal direction, at $F_1 = (-c, 0)$ and $F_2 = (c, 0)$. We call S the set of points for which the sum of the distances to F_1 and F_2 is a fixed constant.

In our image, we have also drawn an ellipse representing the set S around the two foci. The ellipse intersects each coordinate axis in two points, which we have labeled as well. As before, we call the points that lie furthest from the center the **major vertices**, and we call the two lying closest to the center the **minor vertices**. Finally, we have labeled a point P that lies in S.

We know that the sum of the two distances from P to each focus should be a fixed constant. In order to determine what this constant is, we will consider when P lies in a specific position, namely, when $P = (a, 0)$.

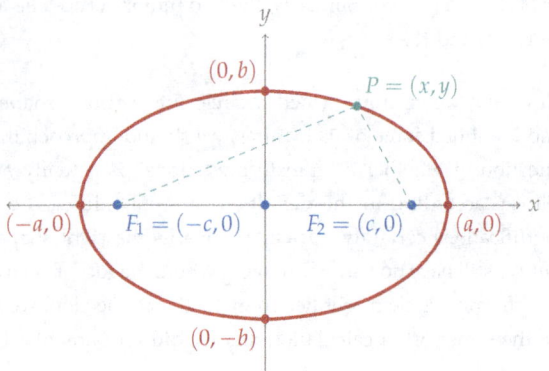
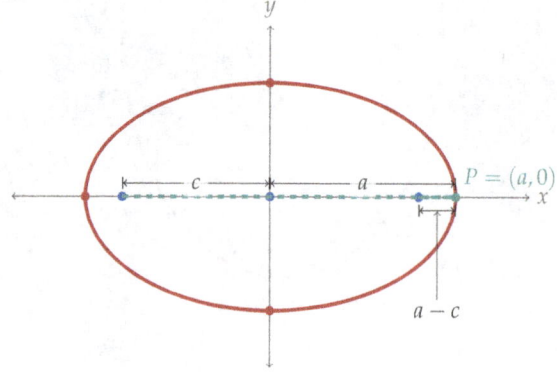

Now we can see that the distance from P to the further focus is $\text{dis}(P, F_1) = a + c$, and the distance from P to the nearer focus is $\text{dis}(P, F_2) = a - c$. Therefore the sum of the distances is

$$\text{dis}(P, F_1) + \text{dis}(P, F_2) = (a+c) + (a-c) = 2a.$$

Next we will position P at one of the minor vertices, $P = (0, b)$. At this point the distances from P to each focus are equal. Since we already know that the sum of the distances is equal to $2a$, we conclude that the distance from $(0, b)$ to each focus must be a.

Because the major and minor axes are perpendicular, the line segments joining the center, focus $(c, 0)$, and minor vertex $(0, b)$ form a right triangle that is interior to the ellipse. We now know that this triangle has sides of length a, b, and c. By the Pythagorean Theorem, these satisfy the equation

$$b^2 + c^2 = a^2.$$

This relationship allows us to identify any one of these lengths, given the other two. In other words, we need only know the center, one focus, and one vertex to determine all other features of an ellipse.

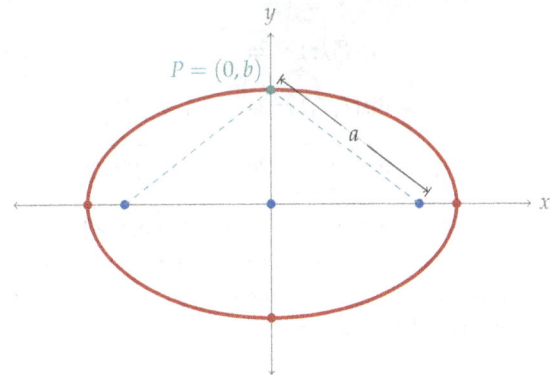

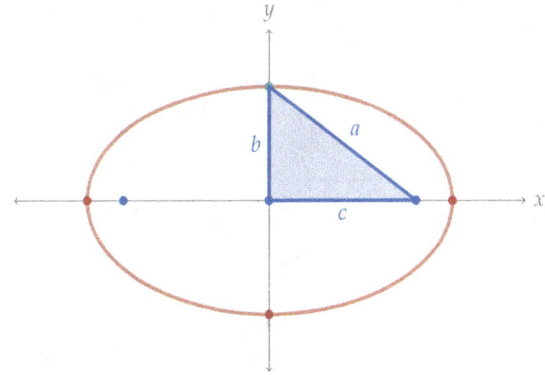

Lesson 19

We now know quite a bit about the shape of an ellipse, and we can use what we know to define all of its major features.

Definition: The line passing through both foci of an ellipse is called its **major axis**. The major axis intersects the ellipse at two points, called the **major vertices**. The point on the major axis lying equidistant from the two foci is called the **center** of the ellipse.

The line that passes through the center and is perpendicular to the major axis is called the **minor axis**. It also intersects the ellipse at two points, called the **minor vertices**.

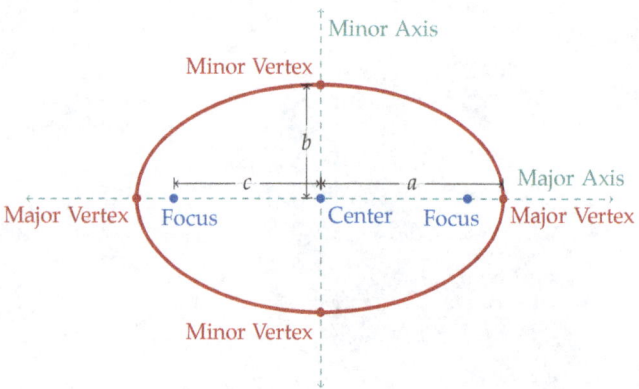

Now we again take $P = (x, y)$ to be any point in S. The distance between P and F_1 can be written

$$\text{dis}(P, F_1) = \sqrt{(x - (-c))^2 + (y - 0)^2}.$$

Likewise, the distance between P and F_2 can be written

$$\text{dis}(P, F_2) = \sqrt{(x - c)^2 + (y - 0)^2}.$$

From our previous work we know that $2a$ is the maximum distance between two points on the ellipse, the distance between the major vertices, and that this is precisely the fixed constant referred to in Apollonius' definition. The sum of the two distances should equal this constant:

$$\text{dis}(P, F_1) + \text{dis}(P, F_2) = 2a$$

$$\text{dis}(P, F_1) = 2a - \text{dis}(P, F_2)$$

$$\sqrt{(x + c)^2 + y^2} = 2a - \sqrt{(x - c)^2 + y^2}.$$

To simplify this equation, we will square both sides. This will eliminate one of the square roots, but the expression on the right-hand side contains more than one term and must be distributed when it is squared.

LESSON 19

This will result in the three expressions that we find enclosed in square brackets in the equation below. We then simplify as much as we are able, until we isolate the remaining square root.

$$(x+c)^2 + y^2 = [4a^2] - \left[2(2a)\sqrt{(x-c)^2 + y^2}\right] + [(x-c)^2 + y^2]$$

$$x^2 + 2cx + c^2 = 4a^2 - 4a\sqrt{(x-c)^2 + y^2} + x^2 - 2cx + c^2$$

$$4a\sqrt{(x-c)^2 + y^2} = 4a^2 - 4cx$$

$$a\sqrt{(x-c)^2 + y^2} = a^2 - cx.$$

At this point we may square both sides again:

$$a^2\left((x-c)^2 + y^2\right) = a^4 - 2(a^2)(cx) + c^2x^2$$

$$a^2\left(x^2 - 2cx + c^2 + y^2\right) = a^4 - 2a^2cx + c^2x^2$$

$$a^2x^2 + a^2c^2 + a^2y^2 = a^4 + c^2x^2$$

$$a^2x^2 - c^2x^2 + a^2y^2 = a^4 - a^2c^2$$

$$(a^2 - c^2)x^2 + a^2y^2 = a^2(a^2 - c^2).$$

Now we recall that the distances a, b, and c satisfy the Pythagorean Theorem. Since a is the longest of the three sides, we have $a^2 = b^2 + c^2$, or alternatively, $a^2 - c^2 = b^2$. Making this substitution, we arrive at:

$$(b^2)x^2 + a^2y^2 = a^2(b^2)$$

$$\frac{x^2}{a^2} + \frac{y^2}{b^2} = 1.$$

What we find is the standard form of a transformed circle, centered at the origin. Therefore, any point (x, y) satisfying Apollonius' definition of the ellipse also satisfies the equation of this transformed circle, and vice versa.

We can use this standard form to identify all the relevant features of an ellipse. In particular, identifying the values of a and b tell us the distances to our major and minor vertices. These ellipses may also be **translated** to any position in a coordinate system, so we should expect an ellipse centered at (h, k) to have equation

$$\frac{(x-h)^2}{a^2} + \frac{(y-k)^2}{b^2} = 1.$$

We'll demonstrate the use of this form with a few examples.

Lesson 19

Example: Find the center, the axes of symmetry, and the vertices of the ellipse:

$$9y^2 + 4x^2 - 18y + 8x - 23 = 0$$

Here we are given a quadratic equation that is not in any standard form, and at first glance we may not even know what kind of conic section this equation represents. We do see two quadratic terms, however, so this equation is not likely to represent a parabola. We also see that the two quadratic terms have the same sign, indicating that the graph of this equation will be an ellipse. In any case, we'll complete the square on both variables to put this equation into standard form.

$$4x^2 + 8x + 9y^2 - 18y = 23$$
$$4\left(x^2 + 2x\right) + 9\left(y^2 - 2y\right) = 23$$
$$4\left(x^2 + 2x + 1\right) + 9\left(y^2 - 2y + 1\right) = 23 + 4 + 9$$
$$4(x+1)^2 + 9(y-1)^2 = 36$$
$$\frac{4(x+1)^2}{36} + \frac{9(y-1)^2}{36} = 1$$
$$\frac{(x+1)^2}{9} + \frac{(y-1)^2}{4} = 1.$$

In this form we can see all of the transformations of a circle that led to this ellipse. We can see that this ellipse has been shifted horizontally and vertically to have its center at $(-1, 1)$. The factor of 9 located in the denominator of the x-term tells us that $a = 3$, and we know that this will be the distance from the center to the horizontal vertices, placing them at $(-4, 1)$ and $(2, 1)$. These are our major vertices. Likewise, from the y-term we find that $b = 2$, placing our minor vertices at $(-1, -1)$ and $(-1, 3)$.

As we know, the major axis passes through the two major vertices. This horizontal line has equation $y = 1$. The minor axis, which passes through the minor vertices, has equation $x = -1$. Knowing all of this, we can draw our ellipse below.

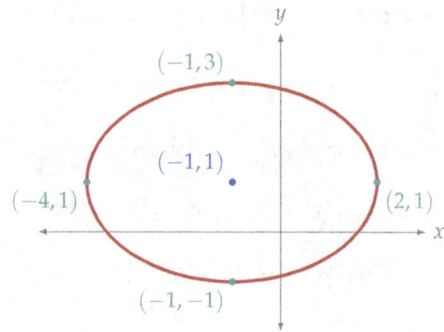

LESSON 19

The next example is the converse of the previous one. Here we are given pieces of information about an ellipse, then asked to find its equation.

Example: An ellipse has its center at $(-1, 4)$. The distance from the center to one of the major vertices is 5 units, and one focus is located at $(2, 4)$. Find its equation.

The key connection that we want to make here involves the coordinates of the two points given us. The fact that they both have the same y-coordinate tells us that they both lie on the horizontal line $y = 4$. Since this line passes through the center and focus, it is our major axis. We know that the major vertices lie 5 units away from the center along this line, placing them at $(-6, 4)$ and $(4, 4)$.

The minor axis will pass through the origin and be perpendicular to the major axis, being the vertical line $x = -1$.

We still need to locate the minor vertices. We don't know their distance from the center, but we do know the distances of the major vertices and foci, which in our standard form are a and c, respectively. We remember that these three distances satisfy the Pythagorean Theorem.

Therefore we will be able to find b. The distance a is the largest of the three values, so that is the hypotenuse:

$$a^2 = b^2 + c^2$$
$$(5)^2 = b^2 + (3)^2$$
$$b^2 = 25 - 9$$
$$b = 4.$$

This places our two minor vertices 4 units away from the center along the minor axis, at $(-1, 0)$ and $(-1, 8)$.

We now know everything there is to know about this ellipse: the location of the center $(h, k) = (-1, 4)$, the fact that $a = 5$, and that $b = 4$. We write our standard form:

$$\frac{(x+1)^2}{25} + \frac{(y-4)^2}{16} = 1.$$

This example demonstrates that the information about an ellipse may be given to us in a variety of forms. In any case, there is a certain minimum amount of information needed to uniquely identify an ellipse and its equation. Here we are given the center, the distance to one vertex, and the position of one focus, and this is sufficient.

LESSON 19

On this page we show a table of the two standard forms for the equation of an ellipse. In reality the forms are identical, but the orientation of an ellipse is determined by the magnitudes of a and b relative to each other. If the denominator of the x-term is larger, the ellipse will be wider in the horizontal direction. If the denominator of y-term is the larger of the two, the ellipse will be wider in the vertical direction.

As a final point, note that this description requires that the coefficients to be in the **denominator**. We often need to rewrite the equation of an ellipse to force it into this form. For example, the equation

$$5x^2 + 13y^2 = 1$$

defines an ellipse centered at the origin, but in standard form this would become

$$\frac{x^2}{\left(1/\sqrt{5}\right)^2} + \frac{y^2}{\left(1/\sqrt{13}\right)^2} = 1.$$

Written this way, we can see that the major axis of this ellipse is actually horizontal, and the major vertices lie at

$$\left(-\frac{1}{\sqrt{5}}, 0\right) \text{ and } \left(\frac{1}{\sqrt{5}}, 0\right).$$

Standard Form $(a > b)$	$\frac{(x-h)^2}{a^2} + \frac{(y-k)^2}{b^2} = 1$
Center	(h, k)
Major Axis	$y = k$
Minor Axis	$x = h$
Major Vertices	$(h-a, k)$ and $(h+a, k)$
Minor Vertices	$(h, k-b)$ and $(h, k+b)$
Foci	$(h-c, k)$ and $(h+c, k)$
Relation	$a^2 = b^2 + c^2$

Standard Form $(a < b)$	$\frac{(x-h)^2}{a^2} + \frac{(y-k)^2}{b^2} = 1$
Center	(h, k)
Major Axis	$x = h$
Minor Axis	$y = k$
Major Vertices	$(h, k-b)$ and $(h, k+b)$
Minor Vertices	$(h-a, k)$ and $(h+a, k)$
Foci	$(h, k-c)$ and $(h, k+c)$
Relation	$b^2 = a^2 + c^2$

LESSON 19

EXERCISES

For Exercises 1–6, put each equation in the standard form
$$\frac{(x-h)^2}{a^2} + \frac{(y-k)^2}{b^2} = 1.$$

1. $27x^2 + 9y^2 = 81$
2. $9x^2 + 4y^2 - 18x + 16y = 11$
3. $x^2 + 2x + y^2 + 4y = 9$
4. $4x^2 + y^2 + 24x - 10y = -45$
5. $3x^2 + 2y^2 - 30x = -69$
6. $3x^2 + 7y^2 - 12x - 28y = -19$

Locate the center of each ellipse, and decide if the major axis is horizontal or vertical.

7. $\dfrac{x^2}{9} + \dfrac{y^2}{4} = 1$
8. $\dfrac{(x+3)^2}{37} + \dfrac{y^2}{18} = 1$
9. $\dfrac{(x-2)^2}{22} + \dfrac{(y-4)^2}{78} = 1$
10. $3546x^2 + 1132(y+1)^2 = 1$

For Exercises 11–14, locate the foci of each ellipse.

11. $\dfrac{x^2}{16} + \dfrac{y^2}{1} = 1$
12. $\dfrac{x^2}{10} + \dfrac{y^2}{36} = 1$
13. $\dfrac{x^2}{81} + \dfrac{(y-5)^2}{49} = 1$
14. $\dfrac{(x+3)^2}{25} + \dfrac{y^2}{9} = 1$

The foci of several ellipses are given with the length of one axis of symmetry. Find the equation of each ellipse.

15. $(0, -8)$, $(0, 8)$, major axis has length 20
16. $(-12, 0)$, $(12, 0)$, minor axis has length 10
17. $(5, 4)$, $(-3, 4)$, major axis has length 10
18. $(-5, -1)$, $(3, -1)$, major axis has length 12

For Exercises 19–22, find the center, foci, and the lengths of the major and minor axes. Then sketch the graph.

19. $\dfrac{x^2}{8} + \dfrac{y^2}{16} = 1$
20. $\dfrac{(x-3)^2}{25} + \dfrac{(y+1)^2}{4} = 1$
21. $9x^2 + 16y^2 = 144$
22. $\dfrac{(x+2)^2}{20} + \dfrac{(y+3)^2}{40} = 1$

23. Find the equation of the ellipse having its center at $(3, 1)$, with the major axis being parallel to the y-axis and 12 units long, and the minor axis being $8\sqrt{2}$ units long.

24. A standard ellipse centered at the origin has an **area** given by the formula $A = \pi ab$. Find the area of the ellipse with vertices at $(\pm 6, 0)$ and foci at $(\pm 2\sqrt{5}, 0)$.

LESSON 20

Hyperbolas

If ellipses are the most familiar of the conic sections, it is likely that hyperbolas are the least familiar. The shape of a hyperbola occurs regularly in natural contexts, but is easily mistaken to be "two parabolas." The reality is that hyperbolas are distinct objects with their own set of properties.

From Lesson 17 we recall three ways to define the shape of a hyperbola. Algebraically, we identified it as the graph of a quadratic equation of two variables, in which the quadratic coefficients are nonzero and have opposing signs. This is a very solid definition, but doesn't help our geometric intuition very much. We also saw Apollonius' definition, which is more geometric in nature.

Definition: Let P and Q be two points. A **hyperbola** is the set of points for which the difference of the distances to P and Q is some fixed constant. The points P and Q are called the **foci** of the hyperbola.

As we've done before, we'll want to use this definition to discover the standard form of the equation for a hyperbola, while simultaneously verifying that both definitions actually define the same object. However, we were able to describe the previous conic sections as transformations of some well-known curves, $y = x^2$ and $x^2 + y^2 = 1$. This gave us good intuition about their shapes and features before we even found their standard forms.

LESSON 20

If hyperbolas are less familiar than parabolas and circles, then we'll have no familiar "standard" curve to transform. To help with our intuition we might recall the third definition of a hyperbola: the intersection of a flat plane with a right circular cone.

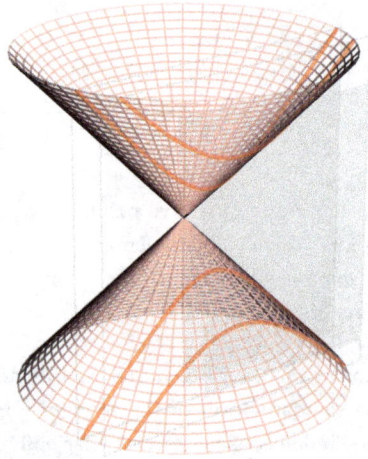

This at least helps us see the general shape of a hyperbola, two of which are shown above. Although, without a coordinate system in place we cannot be very specific about the points or distances in this image.

Therefore we'll again set up a coordinate system in which to place our foci and work with Apollonius' definition. We again try to keep things simple by centering the coordinate system between the foci, placing them at $F_1 = (-c, 0)$ and $F_2 = (c, 0)$. We call S the set of points for which the difference of the distances to F_1 and F_2 is a fixed constant.

Unlike parabolas and ellipses, a hyperbola consists of two distinct branches, and this one never intersects the y-axis. The two points at which the graph intersects the x-axis are called **vertices**, and we label them in our image as $(\pm a, 0)$. The calculation that follows will be nearly identical to the one we did for ellipses, except for a change of sign.

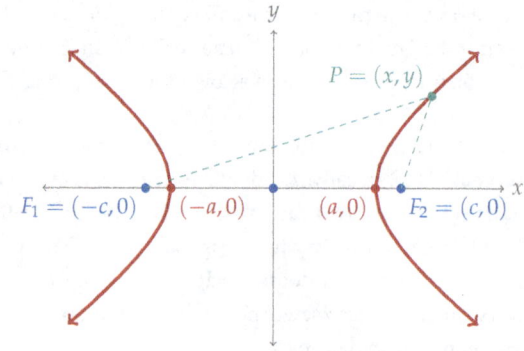

According to Apollonius' definition, if $P = (x,y)$ is a point on S, then the **difference** of the distances from P to each focus should be a fixed constant. To find this constant, we first take P to be one of the two vertices. Specifically, we'll choose the vertex with the positive x-component, so that in our image, the distance to the focus $(-c, 0)$ is the larger of the two: $D(P, F_1) = a + c$. The distance from P to the near focus is $D(P, F_2) = c - a$. Therefore the difference of the two distances is

$$\text{dis}(P, F_1) - \text{dis}(P, F_2) = (a+c) - (c-a) = 2a.$$

This is precisely the same constant that we saw when working with ellipses, and is equal to the total distance between the vertices.

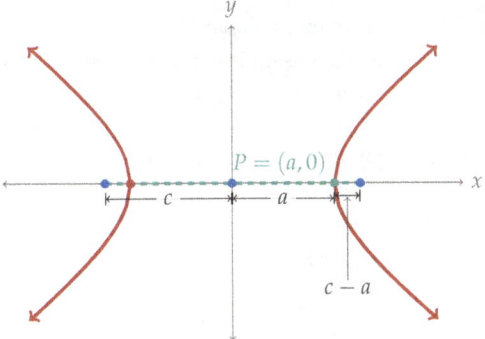

We now have a calculation to perform, in which we take P to be a general point in the set S. To maintain consistency with our images, we'll consider P to be on the right-most branch, having positive x-values, though we could rewrite all of this for P on the left-most branch as well.

With P on the right-most branch, the distance between P and the far vertex F_1 is

$$\text{dis}(P, F_1) = \sqrt{(x - (-c))^2 + (y - 0)^2}.$$

Similarly, the distance between P and F_2 can be written

$$\text{dis}(P, F_2) = \sqrt{(x - c)^2 + (y - 0)^2}.$$

Setting up our equation, we write:

$$\text{dis}(P, F_1) - \text{dis}(P, F_2) = 2a$$

$$\text{dis}(P, F_1) = 2a + \text{dis}(P, F_2)$$

$$\sqrt{(x+c)^2 + y^2} = 2a + \sqrt{(x-c)^2 + y^2}.$$

Recall from our last lesson that when we square both sides of this equation, we will not completely eliminate all the square roots. We must distribute the multiplication on the right-hand side across both of its terms.

LESSON 20

We show in square brackets the three expressions that result from this multiplication. Then, we simplify as much as we are able, until we have the remaining square root isolated on one side.

$$(x+c)^2 + y^2 = [4a^2] + \left[2(2a)\sqrt{(x-c)^2 + y^2}\right] + [(x-c)^2 + y^2]$$

$$x^2 + 2cx + c^2 = 4a^2 + 4a\sqrt{(x-c)^2 + y^2} + x^2 - 2cx + c^2$$

$$4cx - 4a^2 = 4a\sqrt{(x-c)^2 + y^2}$$

$$cx - a^2 = a\sqrt{(x-c)^2 + y^2}.$$

We again square both sides and simplify:

$$c^2x^2 - 2(a^2)(cx) + a^4 = a^2\left(x^2 - 2cx + c^2 + y^2\right)$$

$$a^4 + c^2x^2 = a^2x^2 + a^2c^2 + a^2y^2$$

$$c^2x^2 - a^2x^2 = a^2y^2 + a^2c^2 - a^4$$

$$(c^2 - a^2)x^2 = a^2y^2 + (c^2 - a^2)a^2.$$

When we worked with ellipses, you should recall that at this stage we used the Pythagorean Theorem to make a substitution among the variables a, b, and c. Here we again see the variables a and c representing the same quantities that they did in Lesson 19, the distances to the vertices and foci.

However, in this case we have no minor vertices, and thus no quantity b. To simplify this expression and maintain continuity between our standard forms, we will therefore **define** the quantity b according to the relation

$$b^2 = c^2 - a^2.$$

Now we are free to substitute and finish our simplification.

$$(b^2)x^2 = a^2y^2 + (b^2)a^2$$

$$b^2x^2 - a^2y^2 = b^2a^2$$

$$\frac{x^2}{a^2} - \frac{y^2}{b^2} = 1.$$

Choosing a P on the left-most branch would give this same standard form, an equation that is almost identical to the standard form for an ellipse, apart from a change in sign. Of course, this change in sign is more than just cosmetic. It actually defines an entirely new geometric shape.

However, we began this construction with our foci lying on the x-axis; if we had instead chosen them to lie on the y-axis, we could have gone through this same calculation and arrived at the alternate form

$$\frac{y^2}{b^2} - \frac{x^2}{a^2} = 1.$$

This would define a hyperbola opening **vertically**, in the direction of the y-axis. Thus we can use the signs of the terms in the standard form to learn about the orientation of the hyperbola in the plane.

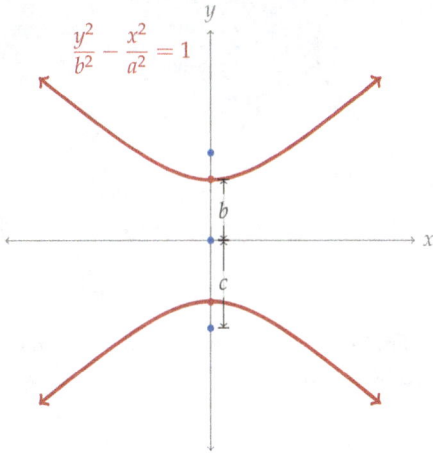

Regardless of orientation, the standard form for the hyperbola involves both of the constants a and b. However, only one of these actually makes a geometric appearance in the graph, the denominator of the positive term, which determines the distance from the center to each vertex.

Like ellipses, however, we still have a relationship between the constants a, b, and c, this time defined by:

$$c^2 = a^2 + b^2.$$

Here the relation is the same for all hyperbolas, as c will always be the longest of the three lengths.

Finally, we note that hyperbolas may be transformed in all the same ways that we transform any graph. Hyperbolas have two axes of symmetry, like ellipses, so this commonly takes the form of a **translation**. An example of this is shown below.

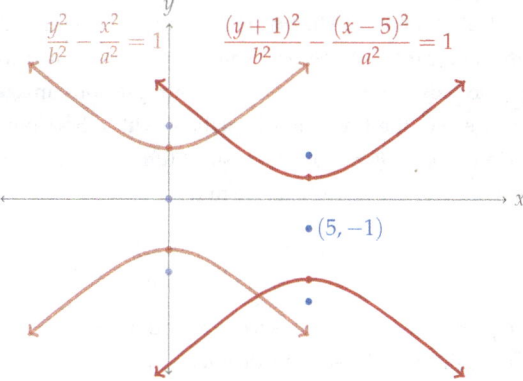

LESSON 20

In our last image we notice that even though the hyperbola has been translated to have a new center, this center lies on a single vertical line with both vertices and both foci. For an ellipse we have called this line the major axis. For a hyperbola, we use a slightly different nomenclature.

Definition: The line passing through both foci of a hyperbola is called its **transverse axis**. The transverse axis intersects the hyperbola at two points, called the **vertices**. The point on the transverse axis lying equidistant from the two foci is called the **center** of the hyperbola. The line passing through the center perpendicular to the transverse axis is called the **conjugate axis**.

These features are all shown on the image at right. It's worth stating that a hyperbola can never intersect its own conjugate axis. We can see this in our previous image, where we saw the translated hyperbola with a horizontal conjugate axis of $y = -1$. If we substitute this value into the equation of the hyperbola, we find

$$\frac{((-1)+1)^2}{b^2} - \frac{(x-5)^2}{a^2} = 0 - \frac{(x-5)^2}{a^2} = 1.$$

This equation has no solutions for x, regardless of the values of a, or b, since the left-hand side must always be negative.

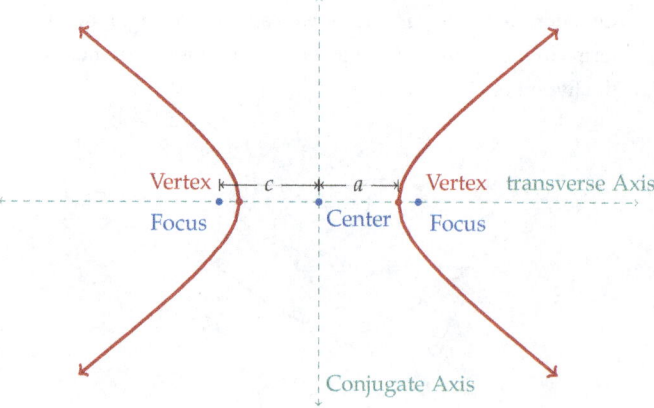

Example: Find the center, vertices, and foci of the hyperbola defined by the equation

$$2y^2 - x^2 + 2x + 8y + 3 = 0.$$

Even if we were not explicitly told that this is a hyperbola, we should be able to recognize it; this equation has two quadratic terms with opposing signs. However, if we did not recognize this right away, we would see it later when it is placed in standard form.

Since this equation is not given in standard form, we will put it into standard form by completing the square:

$$2y^2 - x^2 + 2x + 8y + 3 = 0$$
$$2(y^2 + 4y \quad) - (x^2 - 2x \quad) = -3$$
$$2(y^2 + 4y + 4) - (x^2 - 2x + 1) = -3 + 8 - 1$$
$$2(y+2)^2 - (x-1)^2 = 4$$
$$\frac{(y+2)^2}{2} - \frac{(x-1)^2}{4} = 1.$$

We can now identify our center $(h, k) = (1, -2)$, and since the y-term is positive, we know that our transverse axis is the vertical line $x = 1$. From our denominators, we find the constants $a = 2$ and $b = \sqrt{2}$.

Knowing that the vertices will lie along the vertical transverse axis, we see that b tells us their distance from the center in the vertical direction, placing them at

$$\left(1, -2 - \sqrt{2}\right) \quad \text{and} \quad \left(1, -2 + \sqrt{2}\right).$$

We can also use the Pythagorean Theorem to find the third constant c. With

$$c^2 = a^2 + b^2 = 4 + 2 = 6,$$

we find that $c = \sqrt{6}$. This is the distance of the foci from the center, and since they also lie on the transverse axis, we find the foci at

$$\left(1, -2 - \sqrt{6}\right) \quad \text{and} \quad \left(1, -2 + \sqrt{6}\right).$$

We have just found the five points that were asked for, but if we wanted to draw this hyperbola, we might like to have some information about how **wide** the hyperbola opens to draw a more accurate graph. For ellipses, this information was provided by the minor vertices. For parabolas, we had no such vertices and we relied on another feature to help. We can use this same feature here for hyperbolas.

Definition: A **focal chord** (or **latus rectum**) of a hyperbola is a line segment that passes through one focus, is parallel to the conjugate axis, and has its endpoints on the hyperbola itself. A hyperbola has two focal chords.

In our last example, we determined that the foci lie at

$$\left(1, -2 - \sqrt{6}\right) \quad \text{and} \quad \left(1, -2 + \sqrt{6}\right).$$

If the focal chords are parallel to the conjugate axis, and thus perpendicular to the transverse axis, then they must be horizontal line segments passing through these two points.

Lesson 20

The horizontal line $y = -2 - \sqrt{6}$ represents the lower of the two horizontal lines, with $y = -2 - \sqrt{6} \approx -4.45$. This line will intersect the hyperbola in two places, and we can find their x-coordinates by substitution:

$$\frac{\left((-2-\sqrt{6})+2\right)^2}{2} - \frac{(x-1)^2}{4} = 1$$

$$\frac{\left(-\sqrt{6}\right)^2}{2} - \frac{(x-1)^2}{4} = 1.$$

Note that the negative sign within the parentheses will cancel with itself when squared, so it does not contribute to this calculation. We could have also begun with the line $y = -2 + \sqrt{6}$, and we would arrive at the same solutions.

We continue to simplify:

$$3 - \frac{(x-1)^2}{4} = 1$$

$$\frac{(x-1)^2}{4} = 2$$

$$(x-1)^2 = 8$$

$$x - 1 = \pm\sqrt{8}$$

$$x = 1 \pm \sqrt{8}.$$

Therefore the endpoints of this focal chord will lie at

$$\left(1 - \sqrt{8}, -2 - \sqrt{6}\right) \quad \text{and} \quad \left(1 + \sqrt{8}, -2 - \sqrt{6}\right).$$

As we mentioned earlier, the endpoints of the other focal chord will lie in these same horizontal positions, so we can locate these as well:

$$\left(1 - \sqrt{8}, -2 + \sqrt{6}\right) \quad \text{and} \quad \left(1 + \sqrt{8}, -2 + \sqrt{6}\right).$$

You'll notice that all of the points we have just found have coordinates that are **irrational**. This is normal. Because a conic section is defined by a quadratic equation, it would be unreasonable to expect all of its features to have integer coordinates, or even rational coordinates. For graphing purposes, we can estimate the coordinates of these four endpoints using decimal approximations:

$$(-1.83, 0.45) \quad \text{and} \quad (3.83, 0.45),$$

$$(-1.83, -4.45) \quad \text{and} \quad (3.83, -4.45).$$

Again, the first set of points are the endpoints of the upper focal chord, and the second set are the endpoints of the lower chord. These show use how wide the hyperbola opens horizontally near the foci. We now have a good collection of points with which to create an accurate graph.

LESSON 20

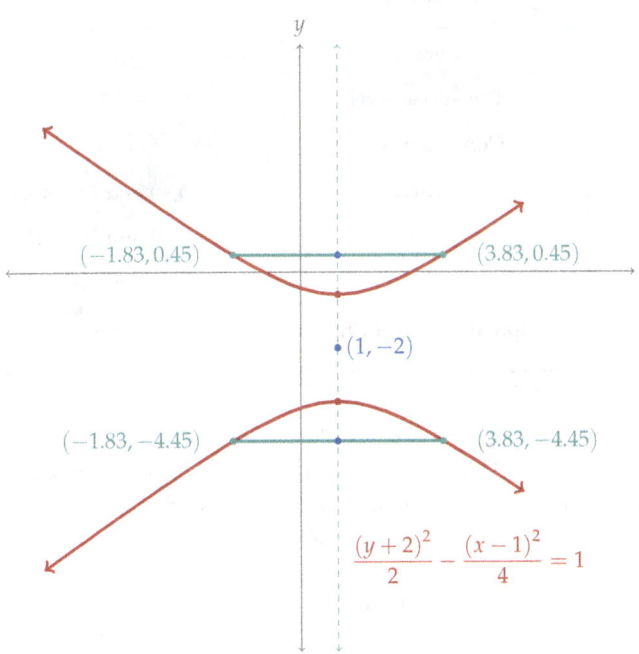

If finding the endpoints of the focal chord is an important part of graphing a hyperbola, then we should make it our next goal to find a formula for their coordinates in terms of the standard form. Since we have just worked with a hyperbola that opens vertically, we will perform this next calculation for a hyperbola that opens **horizontally**.

The standard form of such a hyperbola is

$$\frac{(x-h)^2}{a^2} - \frac{(y-k)^2}{b^2} = 1.$$

With a positive x-term, this hyperbola opens in the direction of that axis, and has its center at (h, k). We are aware that the two foci lie at $(h \pm c, k)$, which are each c units away from the center, along the horizontal transverse axis. The two focal chords are then portions of the vertical lines $x = h \pm c$. We'll find out where these vertical lines intersect the hyperbola by substitution:

$$\frac{((h \pm c) - h)^2}{a^2} - \frac{(y-k)^2}{b^2} = 1$$

$$\frac{(\pm c)^2}{a^2} - \frac{(y-k)^2}{b^2} = 1.$$

We'll solve this equation for y. Just like our previous example, we see that the sign of c is irrelevant to the calculation.

229

Lesson 20

We work to isolate the y-term, and then we find a common denominator for the constant expression:

$$\frac{(y-k)^2}{b^2} = \frac{c^2}{a^2} - 1$$

$$\frac{(y-k)^2}{b^2} = \frac{c^2 - a^2}{a^2}.$$

Recalling that a, b, and c satisfy the relation $c^2 = a^2 + b^2$, we substitute for the numerator on the right-hand side:

$$\frac{(y-k)^2}{b^2} = \frac{(b^2)}{a^2}$$

$$(y-k)^2 = \frac{b^4}{a^2}$$

$$y - k = \pm \frac{b^2}{a}$$

$$y = k \pm \frac{b^2}{a}.$$

Therefore the four endpoints of the two focal chords lie at

$$\left(h \pm c, \, k \pm \frac{b^2}{a} \right).$$

For a hyperbola opening vertically, the calculation is nearly identical, and we leave it to you to verify that formula against the examples we have already done.

Standard Form	$\dfrac{(x-h)^2}{a^2} - \dfrac{(y-k)^2}{b^2} = 1$
Center	(h, k)
Transverse Axis	$y = k$
Conjugate Axis	$x = h$
Vertices	$(h-a, k)$ and $(h+a, k)$
Foci	$(h-c, k)$ and $(h+c, k)$
Relation	$c^2 = a^2 + b^2$
Endpoints of Focal Chords	$\left(h \pm c, \, k \pm \dfrac{b^2}{a} \right)$

In these pages we collect all the features of a hyperbola that we have identified. We will close this lesson by showing how to take information about these features and reconstruct the equation of a hyperbola.

Example: Find the equation of the hyperbola with center at $(4, -1)$, a focus at $(7, -1)$, and a vertex at $(6, -1)$.

Since all of these points have the same y-coordinate, it is simple to identify our horizontal transverse axis as $y = -1$.

LESSON 20

Standard Form	$\dfrac{(y-k)^2}{b^2} - \dfrac{(x-h)^2}{a^2} = 1$
Center	(h,k)
Transverse Axis	$x = h$
Conjugate Axis	$y = k$
Vertices	$(h, k-b)$ and $(h, k+b)$
Foci	$(h, k-c)$ and $(h, k+c)$
Relation	$c^2 = a^2 + b^2$
Endpoints of Focal Chords	$\left(h \pm \dfrac{a^2}{b},\ k \pm c\right)$

We also see that the distance between the center and focus is $c = 3$, and the distance between the center and vertex is $a = 2$. From this we can discover the value of b:

$$c^2 = a^2 + b^2$$
$$9 = 4 + b^2$$
$$b^2 = 5$$
$$b = \sqrt{5}.$$

Being given the center $(h, k) = (4, -1)$, we can fill in all the pieces of our standard form:

$$\dfrac{(x-4)^2}{4} - \dfrac{(y+1)^2}{5} = 1.$$

We will draw a sketch of this hyperbola below. For graphing purposes, we can use the formula we found earlier to identify the endpoints of the two focal chords:

$$(1, -3.5) \quad \text{and} \quad (1, 1.5),$$
$$(7, -3.5) \quad \text{and} \quad (7, 1.5).$$

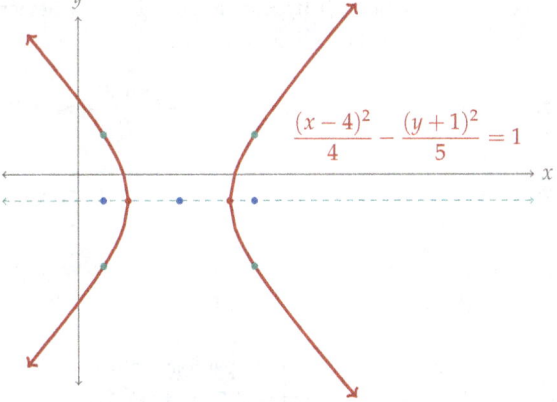

231

LESSON 20

EXERCISES

For Exercises 1–6, put each equation in the standard form
$$\frac{(x-h)^2}{a^2} - \frac{(y-k)^2}{b^2} = 1.$$

1. $5(x-4)^2 - 4(y+2)^2 = 100$
2. $x^2 - 4y^2 + 6x + 16y = 11$
3. $y^2 - 3x^2 + 6y + 6x = 18$
4. $y^2 - 4x^2 - 2y - 16x = -1$
5. $y^2 - 2x^2 - 16 = 0$
6. $9x^2 - 25y^2 + 50y = 18x$

Locate the center of each hyperbola, and decide whether the transverse axis is horizontal or vertical.

7. $\dfrac{x^2}{9} - \dfrac{y^2}{16} = 1$
8. $\dfrac{(y-6)^2}{36} - \dfrac{(x+3)^2}{29} = 1$
9. $\dfrac{(x-7)^2}{2} - \dfrac{y^2}{7} = 1$
10. $(y-9)^2 - (x+5)^2 = 1$

11. Find the equation of the hyperbola with center at $(0,0)$, a focus at $(0,5)$, and a vertex at $(0,3)$.

12. Find the equation of the hyperbola with vertices at $(-2,5)$ and $(12,5)$, and foci at $(-5,5)$ and $(15,5)$.

In Exercises 13–18, find the vertices, foci, and length of the focal chord for each hyperbola. Then sketch its graph.

13. $\dfrac{x^2}{81} - \dfrac{y^2}{36} = 1$
14. $\dfrac{(y-3)^2}{25} - \dfrac{(x-2)^2}{16} = 1$
15. $\dfrac{(x+1)^2}{4} - \dfrac{(y-4)^2}{9} = 1$
16. $y^2 = 36 + x^2$
17. $49x^2 - 16y^2 = 784$
18. $\dfrac{(y+6)^2}{20} - \dfrac{(x-1)^2}{25} = 1$

19. Find the equation of the hyperbola with vertices at $(0,3)$ and $(0,-3)$ that also passes through the point $(2,4)$.

20. Find the equation of the hyperbola with vertices at $(4,0)$ and $(-4,0)$ that also passes through the point $(6,3)$.

21. By plotting points, sketch a graph of the equation
$$\frac{x|x|}{4} - \frac{y|y|}{25} = 1.$$

22. Sketch a graph of the equation $xy = 2$. By changing to a new set of variables with the substitutions $x = s + t$ and $y = s - t$, verify that this also defines a hyperbola.

LESSON 21

Surfaces and Level Curves

From this lesson forward, we will be discussing the graphs of equations in three variables. Just as the set of solutions to an equation of two variables form a one-dimensional curve in $\mathbb{R}^2$, the set of solutions to an equation of three variables will form a two-dimensional **surface** in $\mathbb{R}^3$.

When graphing one-dimensional curves in $\mathbb{R}^2$, it is common to choose values for one of the variables involved and thus determine the value of the other, creating a set of points that lie on the graph. One then infers the rest of the graph from the set of points given. We could use the same strategy here, choosing values for two of the three variables to determine the third.

However, this method will be extremely time-consuming. We will instead use a different but analogous method, in which we set just **one** variable equal to a constant. This will leave us with an equation in two remaining variables, which we can graph in a two-dimensional plane. The collection of these graphs can help us infer the shape of the surface we are looking for.

Definition: The intersection of a curved surface with a flat plane forms a one-dimensional curve that lies completely in that plane, called a **level curve**.

Let's examine some level curves for a few simple surfaces.

LESSON 21

Example: Let S be the surface in $\mathbb{R}^3$ defined by the equation $z = y^2$. Use level curves to sketch the graph of S.

Though the equation given to us has only two variables, we are told that it defines a surface in $\mathbb{R}^3$. We may start by intersecting this surface with vertical planes defined by y, which amounts to setting y equal to different constants.

For example, intersecting the surface with the plane $y = 2$, we find that $z = (2)^2 = 4$. Therefore the set of points lying on both the plane $y = 2$ and the surface $z = y^2$ is

$$Y_2 = \{(x, 2, 4) \mid x \in \mathbb{R}\}.$$

Graphically we can see that this describes a horizontal line extending in the x-direction. This line is generated in three dimensions by the point $(y, z) = (2, 4)$, which lies on the curve $z = y^2$ in the y, z-plane.

Intersecting our surface with any other plane $y = c$ will yield a similar horizontal line in $\mathbb{R}^3$. We can describe each of these lines as a set as we did before, by substituting the equation of the plane $y = c$ into the equation of the surface to find $z = c^2$:

$$Y_c = \{(x, c, c^2) \mid x \in \mathbb{R}\}.$$

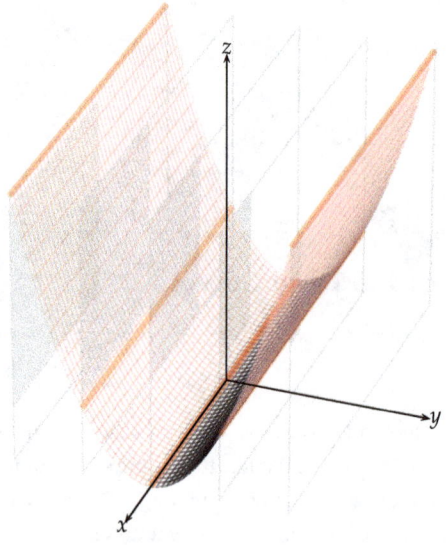

The collection of these lines may give us an indication of our surface, but we can continue with other level curves as well. We'll next turn to level curves defined by z.

If we try setting z equal to a negative constant, then the equation $z = y^2$ cannot have any solutions at all. This means that our surface does not intersect any horizontal planes below the coordinate plane $z = 0$.

For the plane $z = 0$ itself, we find from the equation $z = y^2$ that $y = 0$ as well, so the set of points satisfying both equations is just the x-axis itself:

$$Z_0 = \{(x, 0, 0) \mid x \in \mathbb{R}\}.$$

On the other hand, choosing positive constants c will actually yield two different horizontal lines each, defined by the positive and negative square roots $y = \pm\sqrt{c}$.

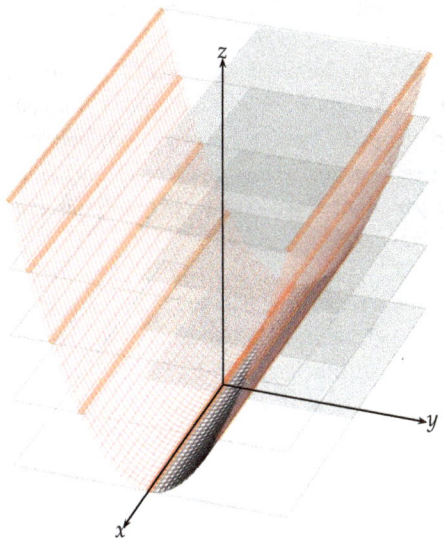

For example, intersecting the S with the horizontal plane $z = 1$ leads to the following sets:

$$Z_1^+ = \{(x, 1, 1) \mid x \in \mathbb{R}\}$$
$$Z_1^- = \{(x, -1, 1) \mid x \in \mathbb{R}\}.$$

This image coincides with what we saw from our previous level curves, so we now have an improved image of S.

Perhaps the most telling option is to use level curves defined by x. Since our equation does not depend on x, all level curves obtained for any choice of x are all identical to each other, defined by the equation $z = y^2$.

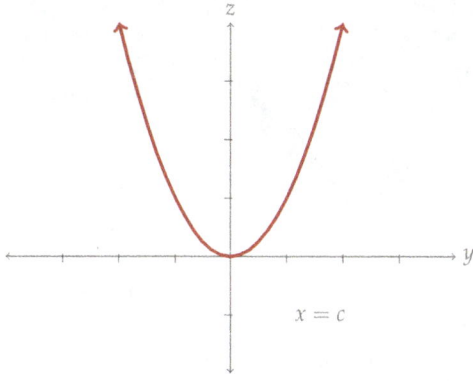

LESSON 21

With an identical level curve for each value of x, drawing many of the planes $x = c$ together in the same three-dimensional space will give us a very clear picture of S.

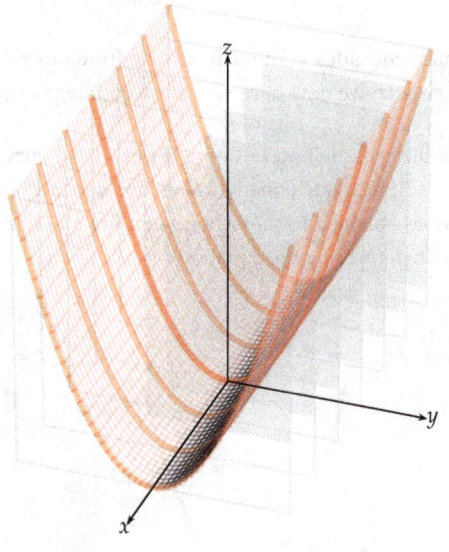

This surface is an example of a **parabolic cylinder**. Its level curves in one variable are all parabolas, and its level curves in its other two variables are lines.

Example: Let S be the surface in $\mathbb{R}^3$ defined by

$$\frac{z^2}{16} - \frac{x^2}{9} = 1.$$

Use level curves to sketch the graph of S.

In this problem we might try an approach similar to the last example, in which we select constant values of x and z and graph the curves that result. With this strategy we would obtain the same sets of level curves that we did previously: straight lines extending in the y-direction.

However, drawing from what we learned from the parabolic cylinder, we recognize the y is the missing variable in this equation, so we'll instead examine the level curves defined by y. These level curves will be identical for any choice of $y = c$, so we might as well choose to graph this equation in the x, z-plane itself.

From Lesson 20, we recognize this as the equation of a hyperbola, already written in standard form. Since z is the positive term in this standard form, the hyperbola has a vertical transverse axis, and opens vertically in that direction. When $x = 0$, we also find that $z^2 = 16$, meaning the hyperbola intercepts the z-axis on this level curve at the points $(0, 0, -4)$ and $(0, 0, 4)$.

We could proceed to find out more about this hyperbola, such as its foci, but for now what we have is sufficient to draw a rough sketch of the level curve. Again, each point on this level curve extends to a straight line in the y-direction to create our surface.

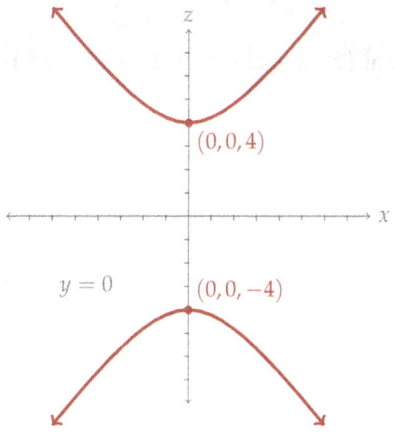

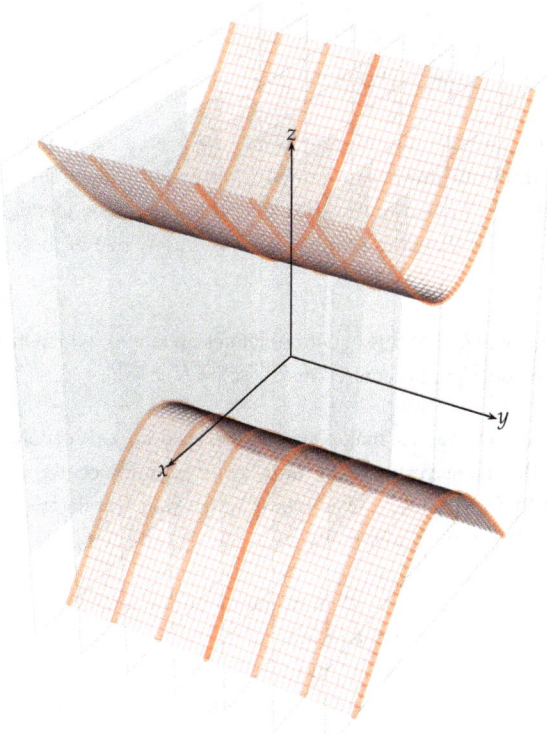

Placing several of the vertical planes $y = c$ together in $\mathbb{R}^3$, we again get a very clear image of S. This surface is an example of a **hyperbolic cylinder**. Its level curves are hyperbolas in one of its variables, and its level curves in each of its other two variables are straight lines.

Lesson 21

Example: Let S be the surface in $\mathbb{R}^3$ defined by

$$\frac{(y-2)^2}{16} + \frac{(z+1)^2}{36} = 1.$$

Use level curves to sketch the graph of S.

In this example the variable x is missing from the equation, and from our previous experience we know that the level curves in all planes defined by that variable will be identical. We can therefore graph this curve in any one of these planes, $x = c$.

We recognize that this equation defines an ellipse, which in our plane $x = c$ is centered at $(x, y, z) = (c, 2, -1)$.

This ellipse has its major axis in the vertical, z-direction, and the major vertices lie 6 units away from the center, at $(c, 2, 5)$ and $(c, 2, -7)$. The minor vertices lie 4 units away from the center, horizontally, at $(c, -2, -1)$ and $(c, 6, -1)$.

It is not necessary to identify the foci if we only wish to make a rough sketch, and indeed we did not find them in our previous examples. But it is worth mentioning that, because the level curve does not depend on the choice of plane $x = c$, the foci of each level curve will lie in an identical location in each plane.

We can find the distance from the center to the two foci using the Pythagorean Theorem:

$$a^2 + c^2 = b^2$$
$$16 + c^2 = 36$$
$$c = \sqrt{20}.$$

The foci lie at $\left(c, -2, \sqrt{20} - 1\right)$ and $\left(c, -2, -\sqrt{20} - 1\right)$.

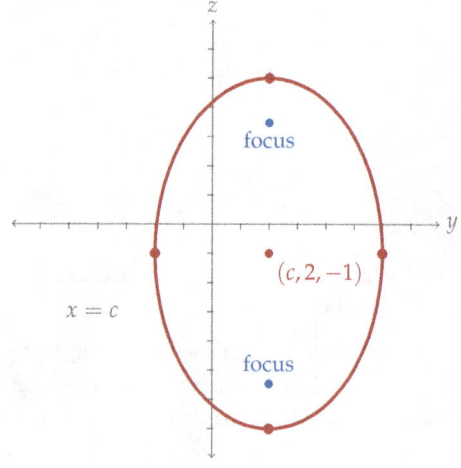

LESSON 21

Drawing several of our level curves in their respective planes, we obtain a very clear impression of our surface. This surface is an example of a **elliptic cylinder**. Its level curves in one variable are ellipses, while its level curves in the remaining two variables are straight lines.

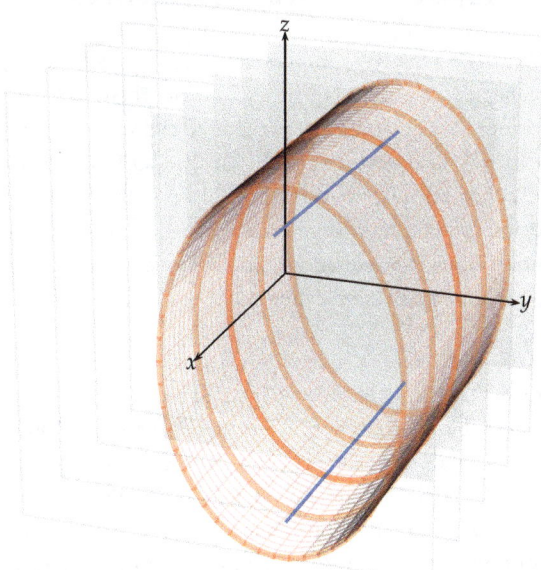

A commonly known special case of the elliptic cylinders is a **circular cylinder**, which has circular level curves.

We may have noticed several facts through these examples:

- Each equation we studied contained only two of three possible variables.

- The variables in each equation had degree two or less.

- The level curves generated by these equations were either straight lines or conic sections.

- The foci of each non-linear level curve lie in an identical location within each constant plane, and thus the set of foci across all constant planes forms a line in $\mathbb{R}^3$, which we call a **focal line**.

In the case of our elliptic cylinder, the foci actually generate **two** lines, as ellipses have two foci each. We cannot write equations for these lines in $\mathbb{R}^3$, but we can write them as sets with a free x-coordinate:

$$F_1 = \left\{ \left(x, -2, \sqrt{20} - 1\right) \mid x \in \mathbb{R} \right\},$$
$$F_2 = \left\{ \left(x, -2, -\sqrt{20} - 1\right) \mid x \in \mathbb{R} \right\}.$$

Lesson 21

More generally, we could create a surface of this type originating with any curve in any plane.

Definition: Let C be a curve lying entirely in a plane, and L be a line that is not contained in the plane. The set of all lines that intersect C and are parallel to L is called a **cylinder**. The curve C is called the **generating curve** of the cylinder, and the parallel lines are called **rulings**.

The generating curves in this lesson have all been conic sections, drawn in one of the three coordinate planes. Also, the rulings have all been parallel to one of the coordinate axes. We saw some less complex examples of cylinders in Lesson 12 when we saw planes, which were also generated by curves in coordinate planes.

More complex cylinders may be generated by still other types of curves, and may be oriented in other directions relative to the coordinate system. These complexities will usually be reflected in the complexity of their equations.

From here we will begin to explore a closely related collection of surfaces. Like the surfaces we have explored in this lesson, they will have conic sections as level curves. But unlike these cylinders, the equations of our new surfaces will involve all three of the variables x, y, and z.

Definition: A **quadratic equation in three variables** is a polynomial equation of the form

$$ax^2 + by^2 + cz^2 + dxy + eyz + fxz + gx + hy + jz - k = 0.$$

The graph of such an equation is a two-dimensional surface called a **quadric surface**. A quadric surface whose center of symmetry lies at the origin is called a **central quadric**.

The planes themselves, though considered cylinders, are not generally considered quadric surfaces. Their equations are **linear**, not quadratic. However, the cylinders we have seen in this lesson are certainly quadric surfaces. Conversely, not all quadric surfaces will be cylinders. The quadrics in this section were each missing one of the three variables from their equations, and this was precisely the property that made them cylindrical.

We will dedicate our remaining lessons to these other quadric surfaces, whose quadratic equations contain all three variables. To simplify our understanding of their graphs, we will focus almost entirely on **central** quadrics. However, we should understand that any central quadric may be shifted to a new position in $\mathbb{R}^3$ using the standard transformations of graphs. We will mention this fact again several times over the next few lessons.

LESSON 21

EXERCISES

In Exercises 1–6, the equations of a surface and a plane are given. Sketch or describe the intersection of the two.

1. $x^2 + y^2 = 1$, $z = 4$
2. $z = y^3$, $x = 2$
3. $z + y^2 = 6$, $z = 2$
4. $x = y^2$, $z + y = 0$
5. $x^2 + z^2 = 4$, $z = y$
6. $x^2 + z^2 = 4$, $z = x$

For Exercises 7–15, describe and sketch each cylinder. Decide whether or not each is a **quadric** surface.

7. $z = 3$
8. $x^2 + y = 4$
9. $y^2 - z^2 = 9$
10. $y = x^4$
11. $3y + 4z = 12$
12. $4x^2 + z^2 = 1$
13. $z - e^x = 0$
14. $y^2 - x = 0$
15. $z - \cos y = 0$

For Exercises 16–21, describe and sketch each surface in $\mathbb{R}^3$.

16. $(z-1)^2 = y + 3$
17. $\dfrac{y^2}{36} + \dfrac{z^2}{25} = 1$
18. $\dfrac{(x-2)^2}{36} + \dfrac{(y+3)^2}{25} = 1$
19. $x^2 - 3(z-2)^2 = 9$
20. $4x^2 + y^2 - 16x - 6y = -21$
21. $z = 3y^2 - 15y + 6$

22. Sketch the solid that is bounded by the four surfaces $z = y^2$, $z = 8 - y^2$, $y = -1$, and $y = 2$.

23. Sketch the solid that is bounded by the three surfaces $4x^2 + 9y^2 = 36$, $z = -1$, and $z + y = 5$.

In Exercises 24–27, identify the focal lines of each quadric surface. Describe these lines as sets of points.

24. $y^2 = -8z$
25. $\dfrac{z^2}{49} + \dfrac{y^2}{9} = 1$
26. $\dfrac{x^2}{16} - \dfrac{y^2}{81} = 1$
27. $y = z^2 - 6z + 11$

In Exercises 28 and 29, a mirror in the shape of a parabolic cylinder is used to reflect sunlight onto a metal pipe that runs along its focal line. This heats a liquid that passes through the pipe, which is then used to produce electricity.

28. If the pipe lies 2 feet above the bottom of the parabolic cylinder, how wide is the cylinder across its focal chord? Write a simple equation that describes the cylinder.

29. If the cylinder is 6ft wide across its focal chord, how far above the bottom of the cylinder should the metal pipe be placed? Write a simple equation that describes the cylinder.

LESSON 22

Ellipsoids and Cones

One common method of creating a quadric surface is to start with a one-dimensional curve in a plane, and then rotate the curve about an axis. A surface generated in this way is known as a **surface of revolution**. Similar to cylinders, we call the original curve the **generating curve**, though in this case the surface is "generated" in a different way.

Most often, we will use a conic section as our generating curve and rotate this conic about one of its axes of symmetry. These surfaces of revolution may be simple to understand geometrically, but we will need to generate such surfaces algebraically as well.

To do so, we will follow a series of steps. Our starting point will always be an equation of two variables, lying in the plane defined by those variables.

- Identify the axis about which we will rotate the curve.
- Substitute the radial coordinate r for the **other** variable.
- Substitute from r back into Cartesian coordinates using the two variables missing from the equation.

We will demonstrate this method using one of the simpler conic sections, an ellipse.

LESSON 22

Example: Generate a surface S by rotating the ellipse

$$\frac{x^2}{9} + \frac{y^2}{16} = 1$$

about the y-axis. Sketch the resulting surface S.

From what we've seen in Lesson 21, this equation alone generates an ellipse in x, y-plane, and when **extended** in the z-direction, and elliptic cylinder in $\mathbb{R}^3$. However, in this instance we are asked to create a different surface by **rotating** the given ellipse into the z-dimension rather than extending it in that direction.

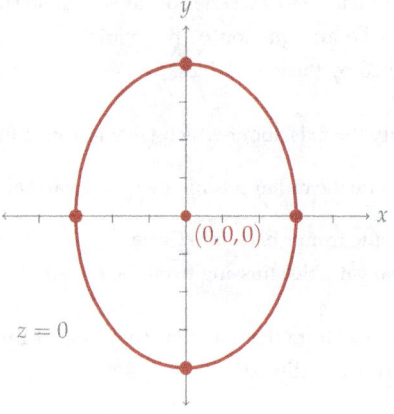

Geometrically, this rotation should give us a sort of oblong spheroid, though we would like to have a more concrete and algebraic description.

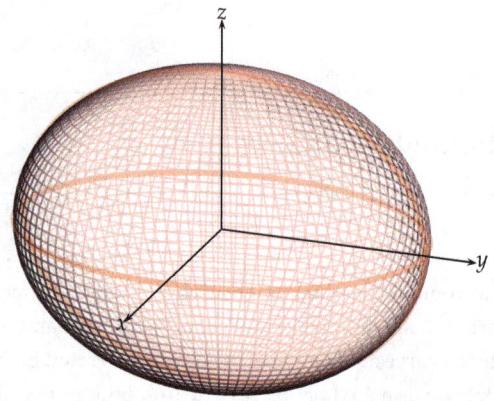

Following the instructions given, we first recognize that we are rotating this ellipse about the y-axis. Therefore we interchange the other variable, x, for the radial coordinate r. This effectively creates an equation in cylindrical coordinates:

$$\frac{(r)^2}{9} + \frac{y^2}{16} = 1.$$

LESSON 22

Next, we use the polar substitution in our remaining variables, $r^2 = x^2 + z^2$, and simplify:

$$\frac{(x^2+z^2)}{9} + \frac{y^2}{16} = 1$$

$$\frac{x^2}{9} + \frac{z^2}{9} + \frac{y^2}{16} = 1$$

$$\frac{x^2}{9} + \frac{y^2}{16} + \frac{z^2}{9} = 1.$$

This now creates a quadratic equation in three variables, whose graph is a quadric surface.

We can look at some level curves of this surface by setting y equal to various constants. For example, if we set $y = 0$:

$$\frac{x^2}{9} + (0) + \frac{z^2}{9} = 1.$$

We can multiply by the common denominator of the x- and z-terms:

$$x^2 + z^2 = 9.$$

This yields a level curve that is a circle of radius 3. If we set y equal to other constants, we will find that all level curves generated by these vertical planes will be circular, precisely because the x- and z-terms have the same denominator.

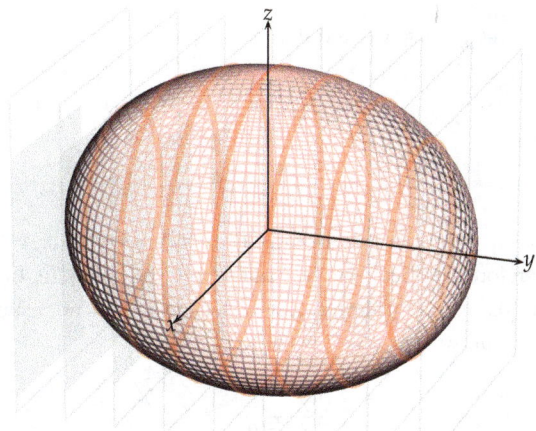

Setting either $y = 1$ or $y = -1$, we find that

$$\frac{x^2}{9} + \frac{z^2}{9} = \frac{15}{16}$$

$$x^2 + z^2 = \frac{135}{16}.$$

These level curves are again circles, this time of radius

$$r = \frac{\sqrt{135}}{4}.$$

245

LESSON 22

If we set $y = \pm 4$, we find that

$$\frac{x^2}{9} + \frac{(\pm 4)^2}{16} + \frac{z^2}{9} = 1$$

$$\frac{x^2}{9} + \frac{z^2}{9} = 0.$$

This equation can only be satisfied if both $x = 0$ and $z = 0$. Therefore our "level curves" are simply the points $(0, 4, 0)$ and $(0, -4, 0)$. Furthermore, if we set y to a constant greater than four, such as $y = 5$, we have

$$\frac{x^2}{9} + \frac{(5)^2}{16} + \frac{z^2}{9} = 1$$

$$\frac{x^2}{9} + \frac{z^2}{9} = -\frac{9}{16}.$$

This equation has no solutions at all, indicating that the plane $y = 5$ does not intersect the surface S.

While the level curves defined by y are each circles, we would find that the level curves defined by x and z will actually be ellipses. The level curve defined by $z = 0$ is the most obvious example, since it is simply the original, unrotated curve,

$$\frac{x^2}{9} + \frac{y^2}{16} + (0) = 1.$$

This particular quadric surface began its life as a surface of revolution, but we can use it to create an new, different quadric surface using graph transformations. For example, if we substitute the expression $z/2$ for the variable z, this should have the effect of stretching our surface by a factor of 2 in the vertical z-direction. Doing so, we find:

$$\frac{x^2}{9} + \frac{y^2}{16} + \frac{z^2}{9} = 1$$

$$\frac{x^2}{9} + \frac{y^2}{16} + \frac{(z/2)^2}{9} = 1$$

$$\frac{x^2}{9} + \frac{y^2}{16} + \left(\frac{z^2}{4}\right)\left(\frac{1}{9}\right) = 1$$

$$\frac{x^2}{9} + \frac{y^2}{16} + \frac{z^2}{36} = 1.$$

Setting $x = y = 0$, we see that this graph intercepts the z-axis at $(0, 0, 6)$ and $(0, 0, -6)$, as expected.

Definition: A quadric surface defined by an equation of the form

$$\frac{x^2}{a^2} + \frac{y^2}{b^2} + \frac{z^2}{c^2} = 1$$

is called an **ellipsoid**. In the case that $a = b = c$, the ellipsoid is actually a **sphere**.

LESSON 22

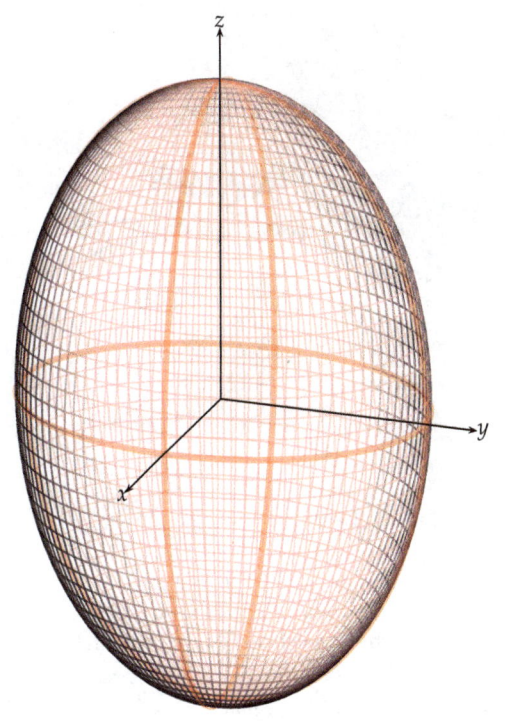

Next, we'll take a look at a surface of rotation that is not generated by a conic section.

Example: Generate a quadric surface S by rotating the line

$$z = \frac{1}{2}y$$

about the z-axis. Sketch the resulting surface S.

Again, although the given equation alone defines a plane in $\mathbb{R}^3$, here we are asked to **rotate** the line into the x-dimension rather than extend it.

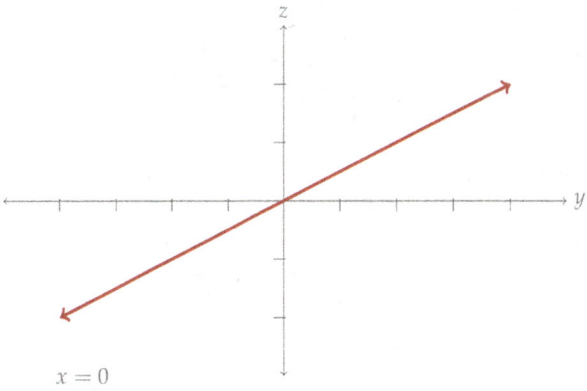

LESSON 22

Therefore, we will go through the standard series of steps. We are asked to rotate around the z-axis, so we replace the other variable y with the radial coordinate r:

$$z = \frac{1}{2}r.$$

Now, we would normally use the polar substitution

$$r^2 = x^2 + y^2,$$

so we will first square both sides of our equation before making the substitution:

$$z^2 = \frac{r^2}{4}$$

$$z^2 = \frac{x^2 + y^2}{4}$$

$$z^2 = \frac{x^2}{4} + \frac{y^2}{4}.$$

With the equation of this quadric surface now in hand, we can look at some level curves. For example, when $z = 0$, we have the equation

$$\frac{x^2}{4} + \frac{y^2}{4} = 0,$$

which has the single solution $(0,0,0)$.

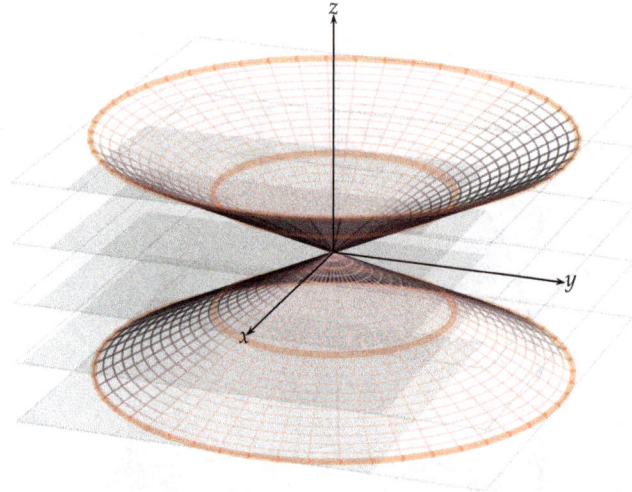

Setting $z = \pm 1$, we obtain the equation

$$x^2 + y^2 = 4,$$

showing that these level curves are both circles of radius 2, lying in the horizontal planes $z = \pm 1$, respectively. Likewise, setting $z = \pm 2$, we obtain the equation

$$x^2 + y^2 = 16.$$

Therefore our level curves in the planes $z = \pm 2$ are circles of radius 4. Because the x- and y-terms have the same denominator, the level curves defined by z in this way will always be perfect circles. This is by construction; after all, this is a surface of revolution.

Turning to the variable x, in the coordinate plane $x = 0$ we have the equation

$$z^2 = (0) + \frac{y^2}{4}$$
$$z = \pm \frac{1}{2} y.$$

This is unsurprising, since this is the equation of our original, unrotated line. The negative part of this equation is the same line after it has been rotated exactly $180°$, to where it once again lies in the y, z-plane.

Next we set $x = \pm 1$. We obtain the equation

$$z^2 = \frac{(\pm 1)^2}{4} + \frac{y^2}{4}$$
$$z^2 - \frac{y^2}{4} = \frac{1}{4}$$
$$4z^2 - y^2 = 1.$$

This equation defines a hyperbola in each of these vertical planes, and since the coefficient of z is positive, these hyperbolas open vertically in the z-direction. Similarly setting $x = \pm 2$ would lead us to the equation

$$z^2 - \frac{y^2}{4} = 1,$$

which is again a hyperbola, distinct from the previous one. The hyperbolas lying in the planes $x = \pm 1$, $x = \pm 2$, and $x = \pm 3$ are all shown in the image below.

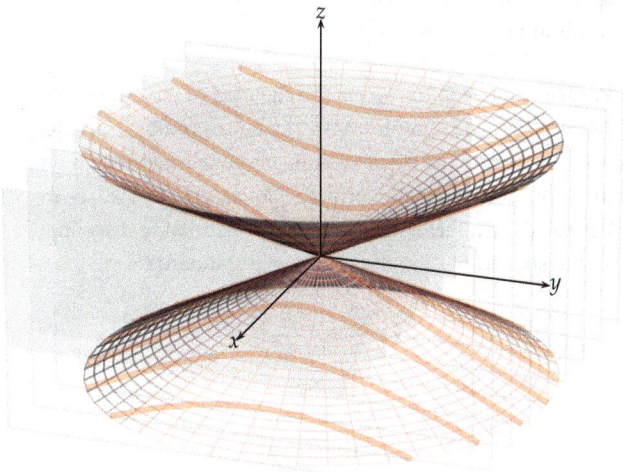

LESSON 22

Actually, what we've just seen is one of the three ways that we **defined** a hyperbola in the first place. A hyperbola is a type of conic section: the intersection of a cone with a plane. However, in Lesson 17 we still lacked an algebraic expression for such a cone, so there was no way to make that definition precise in a coordinate system. We could only make this definition geometrically.

Once we have generated a cone algebraically, as a surface of revolution, we can be precise about intersecting it with specific planes, and obtain equations of the hyperbolas that result in their respective planes.

We are also free to transform the equation of the cone in the usual ways, with graph transformations. In our example, if we substitute the expression $3y$ into our cone's equation in place of the variable y, we should expect our surface to be scaled by a factor of $1/3$ in the horizontal, y-direction. Doing so, and placing the equation in standard form:

$$z^2 = \frac{x^2}{4} + \frac{(3y)^2}{4}$$
$$z^2 = \frac{x^2}{4} + \frac{9y^2}{4}$$
$$z^2 = \frac{x^2}{(2)^2} + \frac{y^2}{(2/3)^2}.$$

Setting $z = \pm 1$ here gives us the equation of an ellipse whose major axis lies in the x-direction, as expected.

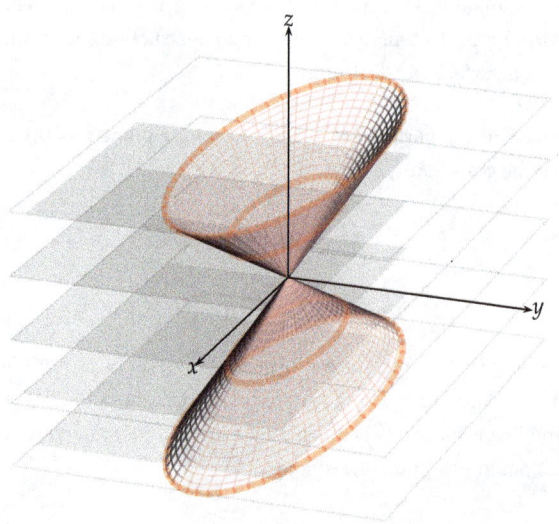

Definition: A quadric surface defined by an equation of the form

$$z^2 = \frac{x^2}{a^2} + \frac{y^2}{b^2}$$

is called an **elliptic cone**. In the case that $a = b$, it is called a **circular cone**.

Here are a few final considerations about the quadric surfaces we've just explored. These are not specific to ellipses or cones, but may be said about any quadratic surface or equation that might define it.

First, the ellipsoids and cones that we've seen in this lesson are all **central** quadric surfaces, meaning that they are centered at the origin. This is not essential. Any quadric surface may be shifted using graph transformations so that it becomes centered at any desired point. For example, the ellipsoid defined by the equation

$$\frac{(x+3)^2}{5} + \frac{(y-2)^2}{7} + \frac{(z+8)^2}{2} = 1$$

is centered at the point $(-3, 2, -8)$. If this is desired, it is important to make the shift transformation **after** making any scaling transformations, as the scalings and shifts may interfere with each other.

Second, throughout this lesson we've used **rotation** about an axis to generate some common quadric surfaces, and later modified them using graph transformations. By their nature, unaltered surfaces of revolution have equations that convert very well into cylindrical coordinates, by simply leaving the radial coordinate r in the equation.

The cone that we graphed earlier,

$$z^2 = \frac{x^2}{4} + \frac{y^2}{4},$$

was earlier written with r substituted for y:

$$z = \frac{1}{2}r.$$

This is often more convenient than writing the equations in Cartesian coordinates.

Finally, we note that not all curves will generate a quadric surface via rotation. As a simple example, we can take the curve $y = x^3$ and rotate it about the x-axis. Our procedure would tell us to change the variable y to the polar coordinate r, giving $r = x^3$, then to square both sides, giving $r^2 = x^6$, and then to substitute:

$$y^2 + z^2 = x^6.$$

But we can see that this equation is **not quadratic**, as the x-term has degree six. The graph of this equation is therefore not a quadric surface.

Conversely, not all quadric surfaces can be generated via a rotation. We will see an example of this in the next lesson.

LESSON 22

EXERCISES

In Exercises 1–4, sketch or describe the intersection of the surface with each of the planes given.

1. $x^2 + y^2 + z^2 = 9$, $z = 0, 1, 2$
2. $100y^2 = 25x^2 + 4z^2$, $y = 1, 2, 3$
3. $y^2 + z^2 - x^2 = 0$, $y = 1, 2, 3$
4. $16x^2 + 9y^2 + 4z^2 = 144$, $x = 0, 1, 2$

Rotating each of the following curves about the given axis creates a surface of revolution. Find the equation of each surface, and determine if each is a **quadric** surface.

5. $z^2 = 4y$, y-axis
6. $z = 3y$, z-axis
7. $16x^2 + 4y^2 = 64$, x-axis
8. $y = \cos z$, z-axis
9. $xy = 2$, x-axis
10. $y^2 = 9z$, y-axis

In each of Exercises 11–14, we are given the equation of a surface of revolution. Identify the generating curve and the axis about which it has been rotated.

11. $x^2 + y^2 - 2z = 0$
12. $4y^2 + 4z^2 = e^x$
13. $x^2 + z^2 = \sin^2 y$
14. $x^2 + y^2 - z^2 = 9$

For Exercises 15–20, identify the graph of each equation as either a cylinder, an ellipsoid, or a cone.

15. $x^2 + 9y^2 + 3z^2 = 1$
16. $x^2 + 9y^2 - 3z^2 = 0$
17. $9y^2 + 3z^2 = 1$
18. $x^2 + 2x + y^2 + 6y = z^2$
19. $7x^4 - 13z^2 = 153$
20. $11x^2 + 17y^2 = 10 - z^2$

For Exercises 21–26, describe and sketch each surface in $\mathbb{R}^3$.

21. $z^2 = \dfrac{x^2}{49} + \dfrac{y^2}{36}$
22. $\dfrac{x^2}{9} + \dfrac{y^2}{25} + \dfrac{z^2}{16} = 1$
23. $3y^2 + 3z^2 - x^2 = 0$
24. $6z^2 - 7x^2 = 42$
25. $x^2 + \dfrac{y^2}{4} + z^2 = 1$
26. $\dfrac{x^2}{9} - \dfrac{y^2}{25} + \dfrac{z^2}{16} = 0$

27. The parabola $z = 4y^2$ in the yz-plane is revolved around the z-axis. Write an equation for the resulting surface using cylindrical coordinates.

28. The hyperbola $2y^2 - z^2 = 2$ in the yz-plane is revolved around the z-axis. Write an equation for the resulting surface using cylindrical coordinates.

LESSON 23

Paraboloids

The next group of quadric surfaces that we approach are the **paraboloids**. Of all the quadric surfaces, paraboloids are perhaps the most widely found in both nature and industry.

There are actually two different classes of paraboloid, and though they have similar equations defining their graphs, their shapes and properties are vastly different. One class is found in applications ranging from billion-dollar telescopes to simple keychain flashlights. The other class has been a favorite in architectural design for decades.

As their name implies, the property that these surfaces have in common is that most of their level curves are parabolas. This cannot be said for the other quadric surfaces. Even though a cone may be intersected with a plane to form a parabola, this plane must be drawn skew to the cone at a very specific angle. As we saw in Lesson 22, the vertical and horizontal level curves are either ellipses or hyperbolas.

We will investigate both types of parabolas, beginning with one generated as a surface of revolution.

LESSON 23

Example: Generate a surface by rotating the parabola

$$z = \frac{1}{9}x^2$$

about the z-axis. Sketch the resulting surface S.

We first perform the rotation about the z-axis. As always, we will perform this rotation by first substituting the polar coordinate r for our other variable, x:

$$z = \frac{1}{9}r^2.$$

As it stands, this is a valid equation for a surface in cylindrical coordinates r, θ, and z. We'll change back to rectangular coordinates to look at the level curves more easily:

$$z = \frac{1}{9}(x^2 + y^2)$$
$$z = \frac{x^2}{9} + \frac{y^2}{9}.$$

You'll notice how similar this seems to the equation of a circular cone from Lesson 22. The difference, of course, is the degree of z, which is linear in the case of a paraboloid, and quadratic in the case of a cone:

$$z = \frac{x^2}{9} + \frac{y^2}{9} \quad \text{versus} \quad z^2 = \frac{x^2}{9} + \frac{y^2}{9}.$$

This is significant because when we look at the level curves in the coordinate planes, such as $x = 0$, we find

$$z = \frac{y^2}{9}.$$

This is the equation of a parabola in the y, z-plane with its vertex at the origin, as opposed to the two straight lines that we saw when graphing the cone.

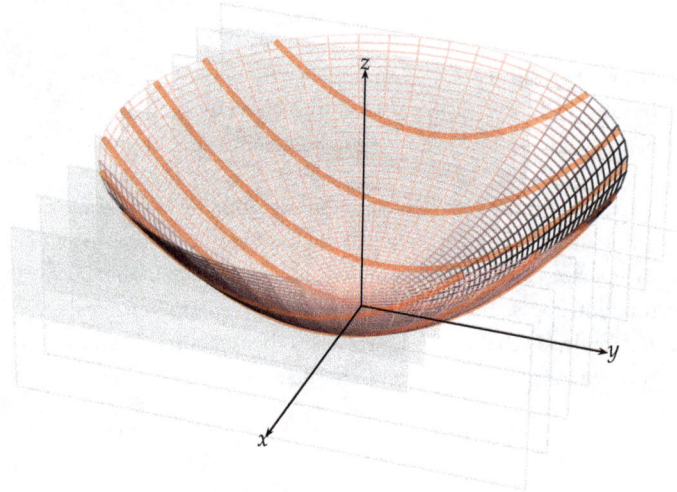

Then, if $x = \pm 1$, we find that the level curves in these planes are also parabolas. However, these have a nonzero constant term that shifts these parabolas in their respective planes, in the positive z-direction:

$$z = \frac{y^2}{9} + \frac{1}{9}.$$

If $x = \pm 2$, we have similar level curves with a larger vertical shift, both lying in the vertical planes $x = 2$ and $x = -2$:

$$z = \frac{y^2}{9} + \frac{4}{9}.$$

Again, if $x = \pm 3$, we find the same parabola, that has a still larger vertical shift, so that the vertices are higher:

$$z = \frac{y^2}{9} + \frac{9}{9} = \frac{y^2}{9} + 1.$$

In the case of the cone, these level curves were not parabolas but were instead **hyperbolas**. We should also note that the numerators of the constant terms are increasing "parabolically" as the magnitude of $x = c$ increases. Of course, this is not a coincidence. We can see why by looking at the other level curves defined by the variable y. Setting $y = 0$:

$$z = \frac{x^2}{9}.$$

This represents a parabola in the x, z-plane, with its vertex at the origin of that coordinate system. This parabola passes precisely through the vertices of the parabolas we have just found. Namely, it passes through the points

$$\left(\pm 1, 0, \frac{1}{9}\right), \left(\pm 2, 0, \frac{4}{9}\right), (\pm 3, 0, 1).$$

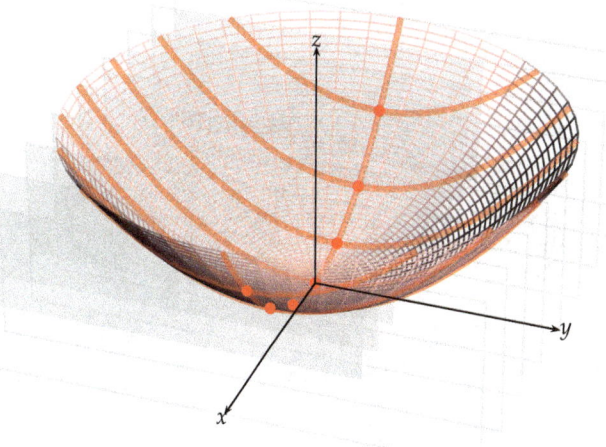

LESSON 23

We can continue finding level curves in the vertical planes defined by y, finding the equations

$$z = \frac{x^2}{9} + \frac{c^2}{9}.$$

The vertices of these parabolas continue to increase in height as the magnitude c increases, exactly as we have seen for the level curves in x. This reflects the symmetry that this equation has in both the x- and y-directions.

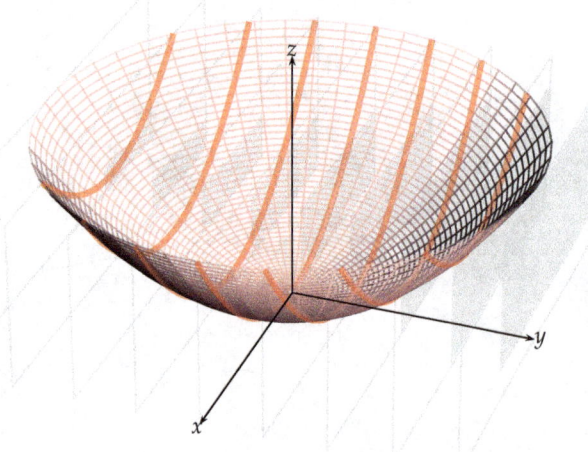

Finally, we can examine the level curves defined by the variable z. We should expect that these level curves will all be circular, since we created this surface with a rotation about that axis. Of course, setting $z = 0$ yields a circle of radius zero, which is simply the point $(0,0,0)$. Setting z equal to a negative number yields no solutions at all.

But in general, so long as c is nonnegative, setting $z = c$ gives us the equation

$$x^2 + y^2 = 9c,$$

which describes a circle of radius $3\sqrt{c}$. Specifically, if $z = 1$, we obtain a circle of radius 3 in the plane $z = 1$.

We can make the relationship between z and the radius of these circles particularly concise by switching back to cylindrical coordinates. When the equation of the paraboloid was written in terms of the radial coordinate,

$$z = \frac{1}{9}r^2,$$

we can see directly the relation between z and the radii of these circular level curves. Solving the equation for r:

$$r = 3\sqrt{z}.$$

LESSON 23

Of course, in this equation the variable r represents the radius of a circle, so it is assumed that we will choose the positive square root. This allows us to find the radius of each level curve by simply choosing a z.

The level curves in the horizontal planes $z = 0$, $z = 1$, $z = 2$, $z = 3$, and $z = 4$ are each shown in the image below.

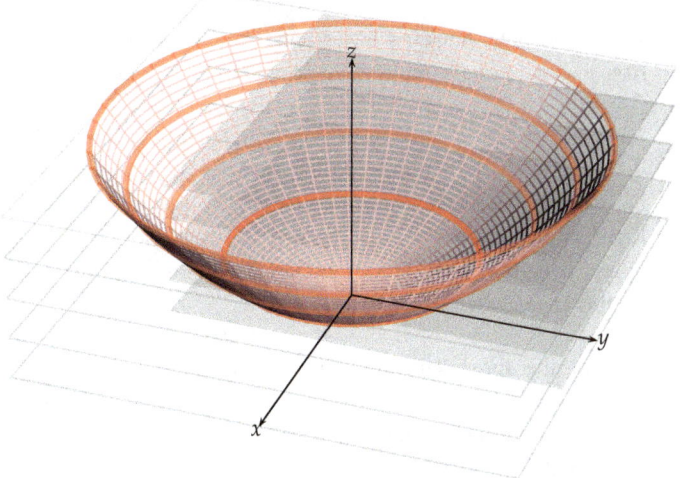

Writing the equation of the paraboloid in cylindrical coordinates can actually help us find one of the defining features of this particular quadric surface. We constructed this surface by rotating a specific parabola about the z-axis, and this original parabola had a **focus** at a point F somewhere on the positive z-axis.

The rotation that we performed will not move this focus; the level curves created by the planes $x = 0$ and $y = 0$ have, respectively, the equations

$$z = \frac{y^2}{9} \quad \text{and} \quad z = \frac{x^2}{9}.$$

Both have their foci lying the same distance away from their common vertex, as will the parabola created by any vertical plane that passes through the origin. Not surprisingly, this common focus is called the **focus of the paraboloid**.

Finding the focus of the paraboloid is not significantly different from finding the focus of any single parabola. We might choose any single level curve for this purpose. Alternatively we might use the equation for the whole paraboloid, written in cylindrical coordinates as

$$z = \frac{r^2}{9}.$$

257

LESSON 23

In any case, we recall from Lesson 17 that to find the distance from the vertex of a parabola to its focus, we should put the parabola's equation into the standard form

$$\pm 4f(z - k) = (r - h)^2.$$

Here, we are changing the variable names slightly to fit our current context, and we are using the variable f to indicate the **focal length**, or distance between the vertex and focus.

Also, in this example our vertex is at the origin, so

$$(h, k) = (0, 0).$$

This means the square for the right-hand side is already complete, and we need only worry about the constants on the left-hand side. Moving the constant to that side:

$$9z = r^2$$

$$4\left(\frac{9}{4}\right)z = r^2.$$

This tells us that the distance from the vertex to the origin is $f = 2.25$, or more specifically, at the point

$$F = \left(0, 0, \frac{9}{4}\right).$$

As usual, paraboloids of this type may be scaled in any dimension using the standard graph transformations. Thus we'll give a more general definition for these quadrics.

Definition: A quadric surface defined by an equation of the form

$$z = \frac{x^2}{a^2} + \frac{y^2}{b^2}$$

is called an **elliptic paraboloid**. In the case that $a = b$, it is called a **circular paraboloid**.

We should note, however, that elliptic paraboloids do not have a single focus in the way that circular paraboloids do.

Now, both the paraboloid from our example and the above definition were written in a specific orientation, with z being a linear variable and the other two being quadratic. In the example, we saw that this resulted in a paraboloid that opens along the z-axis. However, there is no reason to prefer this arrangement.

Any of the three variables may be the linear one, in which case the paraboloid will open along that axis. The paraboloid will open along the positive part of the axis if both quadratic coefficients are positive, and in the negative direction if both are negative.

For example, the image below shows the graph defined by

$$y = \frac{x^2}{16} + \frac{z^2}{4}.$$

For positive values of y, the level curves of this equation are all ellipses with their major axes in the x-direction. The level curves in either x or z are parabolas of different curvatures.

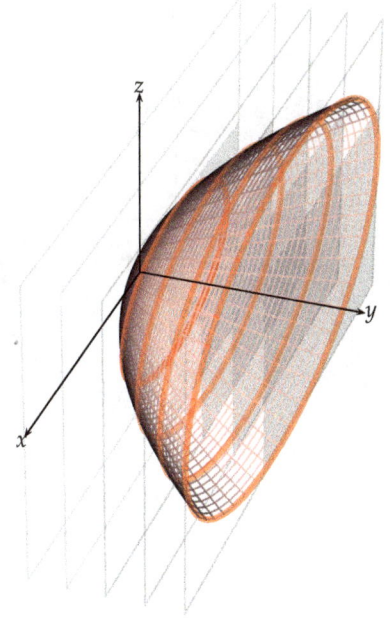

Our strategy of rotating a curve about an axis has served us well thus far, but there is at least one quadric surface that cannot be obtained as a surface of revolution, or even as a transformation of such a surface. The level curves for this surface will not be either circles or ellipses in any orientation. In this instance we will have to take a different approach. Here we give the standard form of the equation first, and then look at the level curves of the equation to try to establish its graph.

Example: Let S be the surface defined by the equation

$$z = \frac{y^2}{9} - \frac{x^2}{4}.$$

Use level curves to sketch the graph of S.

This equation may seem very similar to several we have seen before. Though the right-hand side resembles the hyperbolic cylinder from Lesson 21, the current equation has three variables, not just two. This equation may also resemble the elliptical paraboloids that we have just defined. However, here the signs of the x- and y-terms are different, and this is enough to create an entirely new surface:

$$\frac{y^2}{9} - \frac{x^2}{4} = 1 \quad \text{vs.} \quad z = \frac{y^2}{9} - \frac{x^2}{4} \quad \text{vs.} \quad z = \frac{y^2}{9} + \frac{x^2}{4}.$$

LESSON 23

As mentioned before, we will attempt to sketch this surface by looking at its level curves. This will seem superficially similar to what we just saw with the elliptical paraboloid, though there will be a few key differences with this surface.

For the level curves in x, it is best to start with $x = 0$. In this case we have
$$z = \frac{y^2}{9},$$
which is a parabola opening up in the y, z-coordinate plane, passing through the origin of that coordinate system.

Choosing instead the vertical planes $x = \pm 1$ will give
$$z = \frac{y^2}{9} - \frac{1}{4},$$
which are the same parabolas as for $x = 0$, though these are shifted **down** by 0.25 units. If $x = \pm 2$, we have similar level curves with a larger, negative, vertical shift:
$$z = \frac{y^2}{9} - \frac{4}{4} = \frac{y^2}{9} - 1.$$

When $x = \pm 3$, we find the same parabola, that has a still larger, negative, vertical shift:
$$z = \frac{y^2}{9} - \frac{9}{4}.$$

Similar to we saw with the elliptic parabola, the numerators of the constant terms are decreasing parabolically as the magnitude c increases. However, this is now an inverse relationship, and this is indicated in the image below. When we look at the level curves in y, we will see that this graph does not have the same symmetry between x and y as in the circular paraboloid.

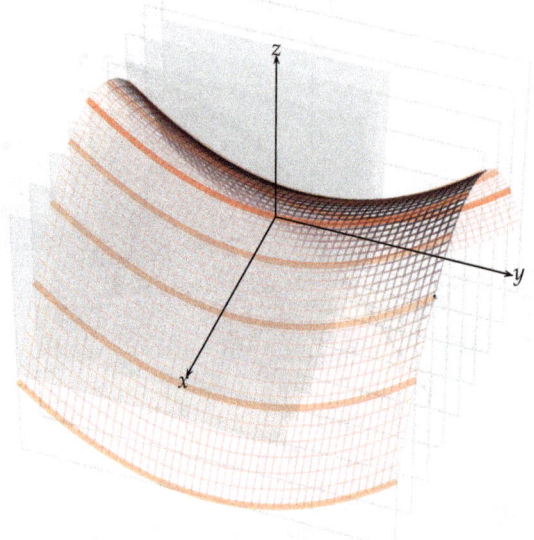

Turning to these level curves in the variable y, we should first set $y = 0$. Doing so will give us the following equation in the x, z-plane, passing through the origin:

$$z = -\frac{x^2}{4}.$$

Contrary to what we saw in the plane $x = 0$ This parabola is facing **down** in the negative z-direction. Again, this parabola passes precisely through each vertex of all the parabolas we found previously. Namely, it passes through the points

$$\left(\pm 1, 0, -\frac{1}{4}\right), (\pm 2, 0, -1), \left(\pm 3, 0, -\frac{9}{4}\right).$$

We next look in the vertical planes $y = 1$ and $y = -1$, in both cases finding the equation

$$z = -\frac{x^2}{4} + \frac{1}{9}.$$

Again, these are both the same parabola as the one defined by $y = 0$, opening in the negative z-direction, but shifted up, in the positive z-direction. The vertices of these two parabolas lie, respectively, at the points

$$\left(0, 1, \frac{1}{9}\right) \quad \text{and} \quad \left(0, -1, \frac{1}{9}\right).$$

Both of these points lie on the previous level curve

$$z = \frac{y^2}{9}$$

that was defined by the plane $x = 0$. All of the points that we've mentioned here are shown on the graph below.

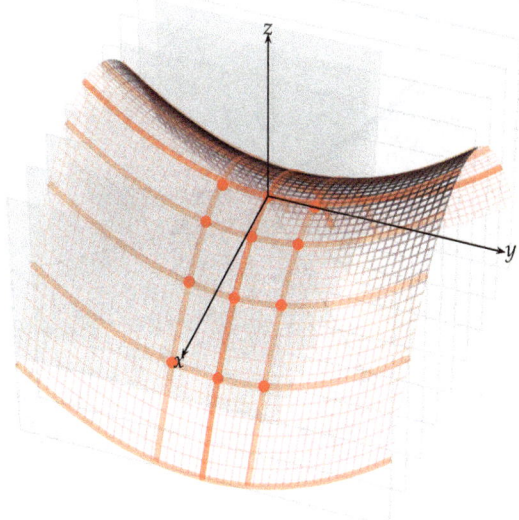

LESSON 23

Other, nonzero values of $y = c$ yield equations

$$z = -\frac{x^2}{4} + \frac{c^2}{9}.$$

These continue to represent downward-facing parabolas, and their vertices will be increasing in height as the magnitude of c increases.

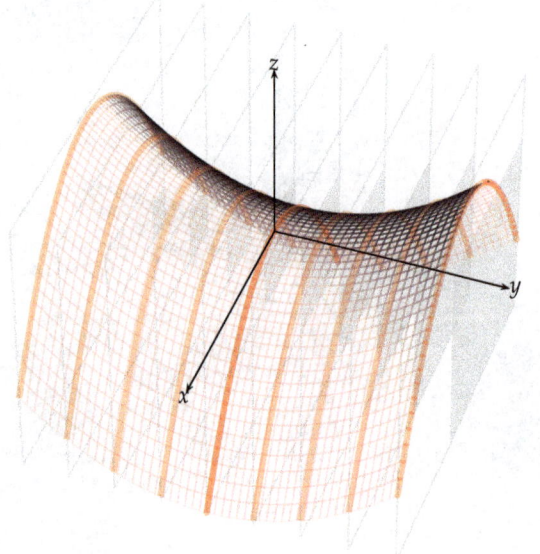

Finally, we consider level curves in z. When $z = 0$, we have

$$\frac{y^2}{9} - \frac{x^2}{4} = 0$$

$$\frac{y^2}{9} = \frac{x^2}{4}$$

$$\frac{y}{3} = \pm\frac{x}{2}$$

$$y = \pm\frac{3}{2}x.$$

So, in the x, y-plane, our level curves are actually two **lines** of different slopes.

Next suppose that $z = 1$. Our equation becomes

$$\frac{y^2}{9} - \frac{x^2}{4} = 1,$$

which describes a **hyperbola**, opening in the y-direction, with its vertices at $(0, \pm 3, 1)$. As the values of c increase from $z = c = 1$, they will yield equations

$$\frac{y^2}{9c} - \frac{x^2}{4c} = 1.$$

These hyperbolas also open in the y-direction. As the positive values $z = c$ increase, the vertices will also increase in distance from the hyperbola's center.

On the other hand, suppose that we choose negative values of z, such as the horizontal plane $z = -1$. Then we have

$$\frac{y^2}{9} - \frac{x^2}{4} = -1$$

$$\frac{x^2}{4} - \frac{y^2}{9} = 1.$$

This hyperbola now opens in the x-direction, with vertices at $(\pm 2, 0, -1)$. For other negative values of z, we'll continue to obtain hyperbolas opening in the x-direction. As the magnitude of $z = c$ increases past $z = -1$, the vertices will increase in distance from the hyperbola's center, similar to what happened with the positive values of z.

The result is a surface that is saddle-shaped, and extends infinitely in all directions. As mentioned earlier, this is not a surface of revolution, nor any transformation of such a surface. It is still a quadric surface, however, as it is defined by a quadratic equation in three variables.

Definition: A quadric surface defined by an equation of the form

$$z = \frac{x^2}{a^2} - \frac{y^2}{b^2}$$

is called a **hyperbolic paraboloid**.

Because the level curves in one variable are hyperboloids, this fact is reflected in the name of these surfaces.

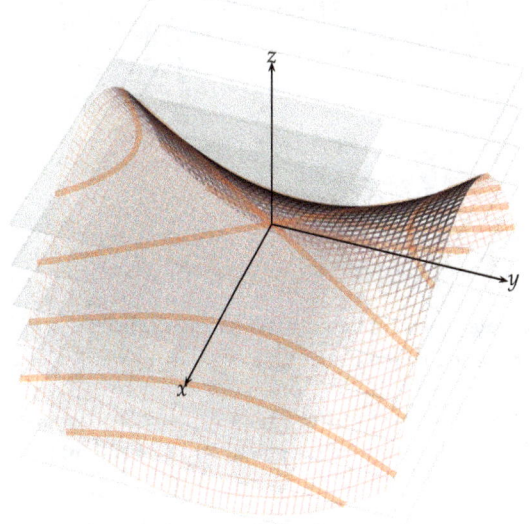

Of course, hyperbolic parabolas may be scaled, shifted, or otherwise transformed just like any other quadric surface. Also, by exchanging the variables x, y, and z, we can change the orientation of the hyperbolic parabola along any axis.

LESSON 23

EXERCISES

In Exercises 1–4, sketch or describe the intersection of the surface with each of the planes given.

1. $x^2 + y - z^2 = 0$, $z = 0, 1, 2$
2. $x = 4y^2 + 6z^2$, $y = -2, 0, 2$,
3. $4z^2 + 9x^2 - y = 0$, $y = 1, 2, 3$
4. $x - 25y^2 - 16x^2$, $x = -1, 0, 1$

For Exercises 5 and 6, consider the graph of $z = \dfrac{x^2}{2} + \dfrac{y^2}{4}$.

5. Locate the foci of the ellipses generated when the surface is intersected by the planes (a) $z = 2$ and (b) $z = 8$.

6. Locate the foci of the parabolas generated when the surface is intersected by the planes (a) $y = 4$ and (b) $x = 2$.

For Exercises 7 and 8, consider the graph of $x = \dfrac{y^2}{9} - \dfrac{z^2}{3}$.

7. Locate the foci of the parabolas generated when the surface is intersected by the planes (a) $y = 3$ and (b) $z = 9$.

8. Locate the foci of the hyperbolas generated when the surface is intersected by the planes (a) $x = -9$ and (b) $x = 3$.

For Exercises 9–14, describe and sketch each surface in $\mathbb{R}^3$.

9. $x^2 - y + z^2 = 0$
10. $x^2 - y^2 + z = 0$
11. $\dfrac{x^2}{36} + \dfrac{y^2}{25} = 1 - \dfrac{z^2}{16}$
12. $9x^2 + 4y^2 - 36z = 0$
13. $z = \sqrt{x + y^2}$
14. $\dfrac{(x-1)^2}{4} + \dfrac{(y+1)^2}{9} = z$

15. Consider a general hyperbolic paraboloid $z = \dfrac{y^2}{b^2} - \dfrac{x^2}{a^2}$. Find its intersection with the plane $ax + by - z = 0$. You may assume that a and b are both positive.

In Exercises 16 and 17, a reflecting telescope has a mirror shaped like a paraboloid of revolution, which reflects all incoming light onto a receiver at its focus.

16. If the distance of the vertex to the focus is 312 inches and the distance across the top of the mirror is 58 inches, how deep is the mirror at its center?

17. If the focus is 1.5 meters from the vertex, and the depth of the mirror is two meters, what is the diameter of the mirror at its opening?

LESSON 24

Hyperboloids

With the exception of the hyperbolic paraboloid, we have so far constructed the bulk of our quadric surfaces by rotating ellipses, lines, or parabolas about an axis. It should be no surprise, then, that our last group of quadric surfaces will be generated from **hyperbolas** in a plane.

Hyperbolas have two axes of symmetry, so we may rotate the curve about either of them to form surfaces of revolution. Both rotations will yield quadric surfaces. However, the two resulting shapes will be very different. One surface will consist of two disconnected "branches" exactly like the original hyperbola. However, we still refer to it as a single surface, just as we refer to a hyperbola as a single curve.

The other rotation will produce a single, unified graph. This surface, because of its unique properties, is widely used in construction and industrial applications.

There are a few subtleties that arise with the level curves of these shapes. In order to not confuse things, and to highlight the differences between these two quadrics and their equations, we begin with a simple, standard hyperbola in the z, y-plane. The transverse and conjugate axes will be the z- and y-axes, respectively, and we will rotate about each of them in turn. Similar to the hyperbolic parabolic from Lesson 23, we will form the equations for the surfaces first, then describe them geometrically using level curves.

LESSON 24

Example: Generate a quadric S by rotating the hyperbola

$$z^2 - y^2 = 1$$

about the z-axis. Sketch the resulting surface S.

Before we begin, let's graph the original hyperbola in the y, z-plane. Setting $y = 0$, we find that the hyperbola intercepts the z-axis at $z = \pm 1$, and opens along that axis. We will rotate about this axis first.

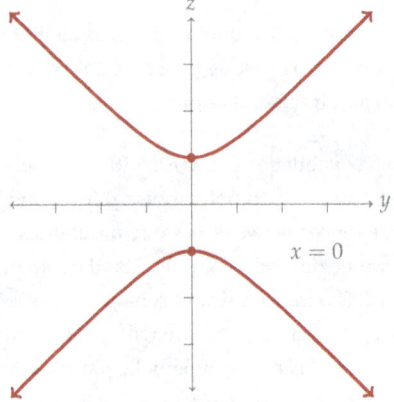

Following our usual procedure, we swap the other variable, y, for the radial coordinate r:

$$z^2 - r^2 = 1.$$

We then change back to Cartesian coordinates in x and y:

$$z^2 - (x^2 + y^2) = 1$$
$$z^2 - x^2 - y^2 = 1.$$

We expect that our level curves in z will be circles, since we rotated the curve about that axis. However, if we set $z = 0$ and simplify, we find

$$-x^2 - y^2 = 1$$
$$x^2 + y^2 = -1.$$

This equation has no solutions. This should not be surprising, because the line $z = 0$ does not intersect our original hyperbola in the y, z-plane.

If we instead choose the horizontal planes $z = \pm 1$, and intersect them with our surface, we find that

$$x^2 + y^2 = (\pm 1)^2 - 1 = 0.$$

These intersections are just the points $(0, 0, 1)$ and $(0, 0, -1)$.

Finally, if we choose horizontal planes $z = c$ for any constant $c > 1$, then all of the equations

$$x^2 + y^2 = c^2 - 1$$

will have solutions, and each will indeed describe a circle. The radii of these circles will increase as the magnitude of c increases. Several of these level curves are shown in the image below, when they exist.

Looking at level curves in the variable x tells a different story. For each constant $x = c$ we will have the equation

$$z^2 - y^2 = c^2 + 1.$$

This defines a hyperbola for any value of c, and all curves open in the same vertical direction. In particular, if $x = 0$ then we have our original hyperbola in the y, z-plane, intersecting the z-axis at $(0, 0, 1)$ and $(0, 0, -1)$.

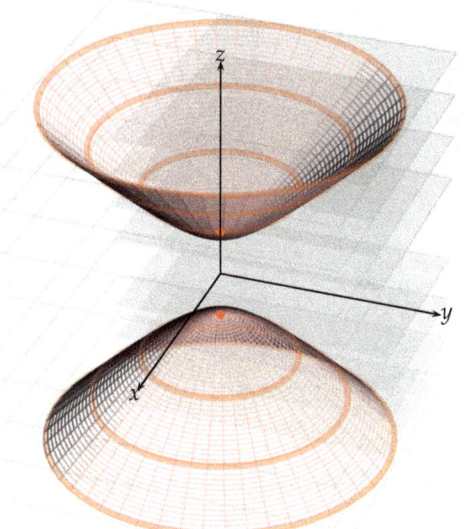
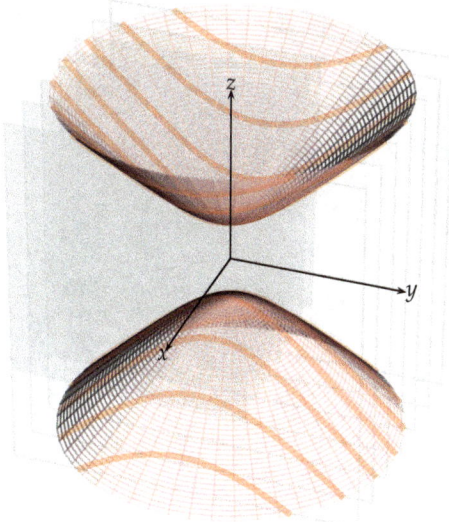

Lesson 24

The situation for the variable y is similar. Intersecting with vertical planes $y = c$ will give

$$z^2 - x^2 = c^2 + 1,$$

again describing hyperbolas opening in the direction of the z-axis. Specifically, if $y = 0$, then the equation $z^2 - x^2 = 1$ describes a hyperbola in the x, z-plane, again intersecting the z-axis at $(0, 0, 1)$ and $(0, 0, -1)$.

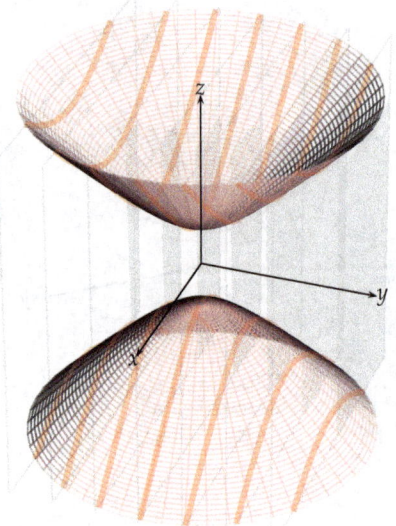

Definition: A quadric surface defined by an equation of the form

$$\frac{z^2}{c^2} - \frac{x^2}{a^2} - \frac{y^2}{b^2} = 1$$

is called an **hyperboloid of two sheets**.

In this definition, we have left the equation in the form we've been using for our example. Changing the signs of the individual terms will alter the orientation of the hyperboloid. For example, the equation

$$\frac{x^2}{a^2} - \frac{y^2}{b^2} - \frac{z^2}{c^2} = 1$$

defines a hyperboloid of two sheets that opens in the direction of the x-axis rather than the z-axis. In this situation, setting $x = 0$ leads to the equation

$$\frac{y^2}{b^2} + \frac{z^2}{c^2} = -1,$$

which has no solutions. So this hyperbola does not intersect the y, z-axis. Similarly, the equation

$$\frac{y^2}{b^2} - \frac{x^2}{a^2} - \frac{z^2}{c^2} = 1$$

defines a hyperboloid of two sheets that opens in the direction of the y-axis, and does not intersect the x, z-plane.

LESSON 24

Let us return to our original hyperbola, $z^2 - y^2 = 1$. Our next task is to rotate this hyperbola about its conjugate y-axis, rather than the z-axis as we have already done.

Example: Generate a quadric S by rotating the hyperbola

$$z^2 - y^2 = 1$$

about the y-axis. Sketch the resulting surface S.

This time we replace the variable z with the radial coordinate r, then change back to rectangular coordinates:

$$r^2 - y^2 = 1$$
$$x^2 + z^2 - y^2 = 1.$$

At first glance this equation may seem to describe an ellipsoid or sphere, but the change of sign in the y-term will change the graph significantly.

It might also be mistaken for the equation of a cone, which we also studied in Lesson 22. The difference here is the presence of a constant on the right-hand side of the equation; the equation of a cone had no constant terms. Again, this is an entirely new surface.

As a surface of revolution, the level curves defined by y will all be circles. Specifically, setting $y = c$ we have

$$x^2 + z^2 = 1 + c^2.$$

We can see that these circles have a minimum radius of 1, which occurs when $c = 0$, and those radii will increase as the magnitude of c increases.

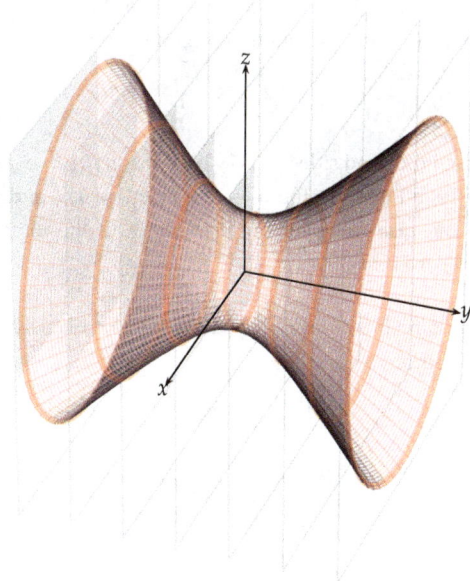

Lesson 24

The level curves for z are a bit more interesting. For horizontal planes $z = c$, our standard form becomes

$$x^2 - y^2 = 1 - c^2.$$

We have three possibilities here. If $c^2 < 1$, then the expression on the right is positive, and this equation describes hyperbolas that open in the x-direction. This is the case for all values of $z = c$ in the open interval $(-1, 1)$.

For $1 < c^2$, the right-hand side is negative. In this case we might rearrange our equation by negating both sides:

$$y^2 - x^2 = c^2 - 1.$$

Since the right-hand side has now been made positive, these hyperbolas all open in the direction of the y-axis, and this will be the case for values of c in either of the intervals $(-\infty, -1)$ or $(1, \infty)$.

The remaining possibility occurs if we set $z = \pm 1$, when the equation becomes

$$x^2 - y^2 = 0$$
$$y^2 = x^2$$
$$y = \pm x.$$

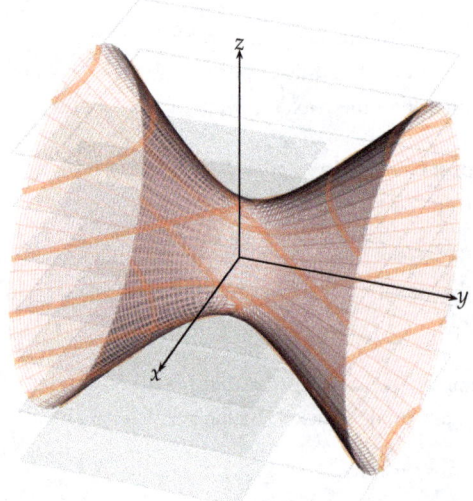

In these particular planes, our level curves are actually two **lines** of slope 1 and -1. All of these level curves are shown, in their respective planes, in the image above.

The level curves in the variable x are similar to those in z. For $x = c$, the standard form becomes

$$z^2 - y^2 = 1 - c^2.$$

We again have three possibilities for these level curves.

270

If $c^2 < 1$, then the right-hand side of the equation is positive, and the level curves are hyperbolas opening vertically in the direction of the z-axis. In particular, for $x = 0$, we again find our original hyperbola in the y, z-plane:

$$z^2 - y^2 = 1.$$

For all constants c in which $1 < c^2$, we can rewrite the equation by negating both sides:

$$y^2 - z^2 = c^2 - 1.$$

Therefore, these hyperbolas will open in the direction of the y-axis. Finally, when $x = \pm 1$, we find the equation $z^2 = y^2$, or $z = \pm y$, which again describes two straight lines of slope 1 and -1 in their respective planes.

Definition: A quadric surface defined by an equation of the form

$$\frac{x^2}{a^2} + \frac{z^2}{c^2} - \frac{y^2}{b^2} = 1$$

is called an **hyperboloid of one sheet**.

Again, these variables may be exchanged, with the single negative term indicating the axis about which the hyperboloid is rotated.

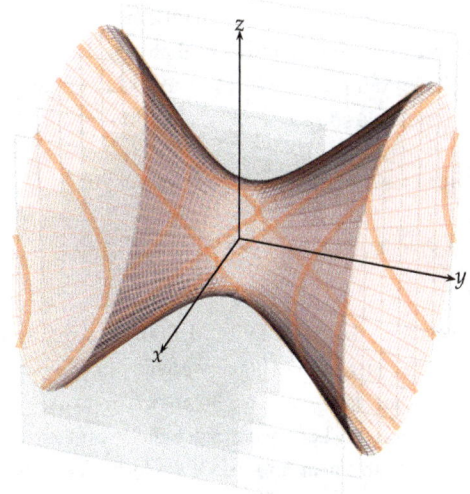

The fact that some of this surface's level curves are straight lines is significant. Like cylinders, cones, and hyperbolic paraboloids which also have lines for level curves, the hyperboloid of one sheet can be constructed using **only** straight lines, if they are arranged in a certain configuration. In architectural design, this allows these curved shapes to be constructed using straight beams. Coupled with the fact that the hyperboloid is structurally very stable, it finds use in many applications.

Lesson 24

Now that we have seen all of the quadric surfaces, their level curves, and the standard forms of their equations, we will look at a few examples in which we are asked to identify an unknown quadric surface from its equation.

Example: Identify the quadric surface defined by

$$9x^2 - y^2 - 4z^2 + 54x + 8z + 41 = 0.$$

Looking back at the collection of definitions we have compiled over the last four lessons, we will put this equation into a standard form and try to match it to one of the quadric surfaces we have seen. Since the x- and z-variables appear with both linear and quadratic terms, we will need to complete the square for those two variables.

$$9x^2 - y^2 - 4z^2 + 54x + 8z + 41 = 0$$
$$9x^2 + 54x - y^2 - 4z^2 + 8z = -41$$
$$9(x^2 + 6x) - y^2 - 4(z^2 - 2z) = -41$$
$$9(x^2 + 6x + 9) - y^2 - 4(z^2 - 2z + 1) = -41 + 81 - 4$$
$$9(x+3)^2 - y^2 - 4(z-1)^2 = 36$$
$$\frac{(x+3)^2}{4} - \frac{y^2}{36} - \frac{(z-1)^2}{9} = 1.$$

Here we see that this has three quadratic variables, and two of them are negative. Because we also have the constant 1 on the right-hand side of the equation, we find that this matches the definition of a **hyperboloid of two sheets**, opening in the direction of the x-axis. This particular hyperboloid is not a central quadric; it been translated so that its center is a $(-3, 0, 1)$.

Example: Identify the quadric surface defined by

$$60z - 20 + 10y^2 - 18x^2 - 20y + 15z^2 = 0.$$

At first glance, we see that this equation has three quadratic variables. This alone tells us that the surface cannot be a cylinder of any type, nor can it be either of the paraboloids. We see that the quadratic terms do not all have the same sign, so we may rule out ellipsoids as well.

Since two of the signs are positive, and the third is negative, and we also have a nonzero constant term involved in the equation, we may want to conclude that this is a hyperboloid of one sheet. However, we have not yet put this equation into its proper, standard form. We do not yet know how the constants will simplify, or if any of the intermediate terms will cancel.

Of course, we will need to complete the square in the y- and z-variables, and we do this here.

$$60z - 20 + 10y^2 - 18x^2 - 20y + 15z^2 = 0$$
$$10y^2 - 20y + 15z^2 + 60z - 18x^2 = 20$$
$$10(y^2 - 2y\ ) + 15(z^2 + 4z\ ) - 18x^2 = 20$$
$$10(y^2 - 2y + 1) + 15(z^2 + 4z + 4) - 18x^2 = 20 + 10 + 60$$
$$10(y-1)^2 + 15(z+2)^2 - 18x^2 = 90$$
$$\frac{(y-1)^2}{9} + \frac{(z+2)^2}{6} - \frac{x^2}{5} = 1.$$

After completing the squares for both y and z, we find that the constant term on the right-hand side of this equation is now zero. Therefore this equation can even be rewritten as

$$x^2 = \frac{5}{9}(y-1)^2 + \frac{5}{6}(z+2)^2,$$

showing that it in fact describes an **elliptic cone**, and not a hyperboloid as we may have expected.

While it may be tempting to forgo the algebra of putting the equation into a standard form, and rely only on the number of quadratic variables and their respective signs, such an approach will not always lead you in the right direction.

Additionally, having the equation in standard form will tell you something about the surface's features on top of simply identifying the type of quadric that the equation defines.

Here we've seen that our example defines an elliptic cone opening in the direction of the x-axis. We can also see that it not a central quadric, but has been translated so that its center (or vertex) lies at $(0, 1, -2)$. It may be difficult or impossible to learn these things from simply looking at the original equation.

Throughout these lessons, **level curves** have been an essential tool for us as we learn about quadric surfaces. But level curves are not useful only for quadric surfaces; they may be used to study any surfaces at all, whether or not they are defined by a quadratic equation. Equations like

$$\left[x^2 + y^2 + z^2 + (3)^2 - (1)^2\right]^2 = 4(3)^2 \left(x^2 + y^2\right)$$

still define perfectly normal surfaces in $\mathbb{R}^3$, and we can use level curves to help describe their shapes. In fact, with a bit of practice and good algebra skills, we can intersect surfaces with planes that are not parallel to any of the coordinate planes, but rather skew in $\mathbb{R}^3$. Choosing planes strategically can tell us even more about our surfaces and their geometry.

LESSON 24

THE QUADRIC SURFACES

$4cy = x^2$

Parabolic Cylinder

$\dfrac{x^2}{a^2} + \dfrac{y^2}{b^2} - \dfrac{z^2}{c^2} = 0$

Elliptic Cone

$\dfrac{x^2}{a^2} + \dfrac{y^2}{b^2} + \dfrac{z^2}{c^2} = 1$

Ellipsoid

$\dfrac{x^2}{a^2} + \dfrac{y^2}{b^2} = 1$

Elliptic Cylinder

$z = \dfrac{x^2}{a^2} + \dfrac{y^2}{b^2}$

Elliptic Paraboloid

$\dfrac{x^2}{a^2} + \dfrac{y^2}{b^2} - \dfrac{z^2}{c^2} = 1$

Hyperboloid of One Sheet

$\dfrac{x^2}{a^2} - \dfrac{y^2}{b^2} = 1$

Hyperbolic Cylinder

$z = \dfrac{x^2}{a^2} - \dfrac{y^2}{b^2}$

Hyperbolic Paraboloid

$\dfrac{x^2}{a^2} - \dfrac{y^2}{b^2} - \dfrac{z^2}{c^2} = 1$

Hyperboloid of Two Sheets

Here we've collected the standard forms of all the quadric surfaces. You'll notice that these are all central quadrics, though any of them may be translated away from the origin using the usual graph transformations. They are also written with a certain orientation in x, y, z-space.

This orientation can be altered as well, by interchanging the variables in each equation. While it is sometimes possible to identify quadric surfaces by the number of quadratic variables present in the equation and their signs, we should always try to put them in one of these standard forms.

LESSON 24

EXERCISES

In Exercises 1–4, sketch or describe the intersection of the surface with each of the planes given.

1. $2x^2 + y^2 - z^2 = 1$, $z = 0, 1, 2$
2. $x^2 - y^2 - 2z^2 = 1$, $y = 0, 1, 2$
3. $4x^2 - 3y^2 + 12z^2 + 12 = 0$, $x = 0, 1, 2$
4. $x^2 + 5y^2 = z^2 - 1$, $z = 1, 2, 3$

For Exercises 5–12, determine the type of quadric surface that is defined by the given equation.

5. $y^2 + 4z^2 - x = 0$
6. $4x^2 + y^2 + 4z^2 - 4y - 24z + 36 = 0$
7. $(x+y)^2 + (x-y)^2 = 1 - 2z^2$
8. $x^2 - y^2 + z^2 - 2x + 2y + 4z + 2 = 0$
9. $x^2 + 2z^2 - y^2 + 4y = 4$
10. $x^2 + 2z^2 = 1$
11. $x^2 - y^2 + z^2 - 4x - 2y - 2z + 7 = 0$
12. $4x^2 - y^2 + 4z^2 = 0$

For Exercises 13–18, describe and sketch each surface in $\mathbb{R}^3$.

13. $z^2 = 9x^2 + 4y^2 + 36$
14. $z^2 - y^2 - \dfrac{x^2}{4} = 1$
15. $\dfrac{x^2}{16} + \dfrac{y^2}{25} - \dfrac{z^2}{9} = 1$
16. $x^2 + y^2 - 16x + z^2 = -64$
17. $(x-1)^2 = 2y^2 + 3z^2$
18. $\dfrac{y^2}{4} = z^2 + \dfrac{x^2}{9} + 1$

For Exercises 19 and 20, consider the graph of $z^2 = \dfrac{x^2}{2} + \dfrac{y^2}{4}$.

19. Locate the foci of the ellipses generated when the surface is intersected by the planes (a) $z = 2$ and (b) $z = 4$.

20. Locate the foci of the hyperbolas generated when the surface is intersected by the planes (a) $y = 4$ and (b) $x = 2$.

21. A cooling tower for a nuclear reactor is constructed in the shape of a hyperboloid of one sheet. The diameter at the base is 300m and the minimum diameter, 500m above the base, is 200m. Find an equation describing the shape of the tower, using coordinates in which the origin is at the center of the narrowest part of the tower.

TABLE OF TRIGONOMETRIC IDENTITIES

The Complementary Angle Theorem

$\sin\theta = \cos(90° - \theta)$ $\cos\theta = \sin(90° - \theta)$

$\csc\theta = \sec(90° - \theta)$ $\sec\theta = \csc(90° - \theta)$

$\tan\theta = \cot(90° - \theta)$ $\cot\theta = \tan(90° - \theta)$

Sum/Difference Formulas

$\sin(\alpha + \beta) = \sin\alpha\cos\beta + \cos\alpha\sin\beta$

$\sin(\alpha - \beta) = \sin\alpha\cos\beta - \cos\alpha\sin\beta$

$\cos(\alpha + \beta) = \cos\alpha\cos\beta - \sin\alpha\sin\beta$

$\cos(\alpha - \beta) = \cos\alpha\cos\beta + \sin\alpha\sin\beta$

$\tan(\alpha + \beta) = \dfrac{\tan\alpha + \tan\beta}{1 - \tan\alpha\tan\beta}$

$\tan(\alpha - \beta) = \dfrac{\tan\alpha - \tan\beta}{1 + \tan\alpha\tan\beta}$

Pythagorean Identities

$\sin^2\theta + \cos^2\theta = 1$

$\tan^2\theta + 1 = \sec^2\theta$

$1 + \cot^2\theta = \csc^2\theta$

Even/Odd Properties

$\sin(-\theta) = -\sin(\theta)$

$\cos(-\theta) = \cos(\theta)$

The Law of Sines

$\dfrac{\sin A}{a} = \dfrac{\sin B}{b} = \dfrac{\sin C}{c}$

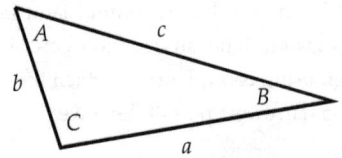

The Law of Cosines

$c^2 = a^2 + b^2 - 2ab\cos C$

$b^2 = a^2 + c^2 - 2ac\cos B$

$a^2 = b^2 + c^2 - 2bc\cos A$

Double-Angle Formulas

$\sin(2\alpha) = 2\sin\alpha\cos\alpha$

$\cos(2\alpha) = \cos^2\alpha - \sin^2\alpha$

$\cos(2\alpha) = 1 - 2\sin^2\alpha$

$\cos(2\alpha) = 2\cos^2\alpha - 1$

$\tan(2\alpha) = \dfrac{2\tan\alpha}{1 - \tan^2\alpha}$

Half-Angle Formulas

$\sin^2\alpha = \dfrac{1}{2}(1 - \cos 2\alpha)$

$\cos^2\alpha = \dfrac{1}{2}(1 + \cos 2\alpha)$

$\sin\left(\dfrac{\alpha}{2}\right) = \pm\sqrt{\dfrac{1 - \cos\alpha}{2}}$

$\cos\left(\dfrac{\alpha}{2}\right) = \pm\sqrt{\dfrac{1 + \cos\alpha}{2}}$

CPSIA information can be obtained
at www.ICGtesting.com
Printed in the USA
LVHW020705100822
725543LV00003B/25